MEN OF
STEEL

MEN OF STEEL

LOUIS A ROSATI
PHOTOGRAPHY BY MICHAEL A ROSATI

CONTENTS

PREFACE

Lockport, New York. It's not what you would have called a "steel town," at least not like those scattered along the Monongahela north and south of Pittsburgh or in Lackawanna, N.Y., where the steel mill was the major or only employer; but Lockport has a metals history dating to the nineteenth century. In the 1940s and 1950s, manufacturing and business in Lockport thrived around a vibrant urban core. Simonds Saw and Steel was the town's second largest employer, its workers dwarfed in number by a factor of six to one to the leading employer, Harrison Radiator Division of General Motors. However, Simonds counted for something, because even though it didn't dominate the industrial or city landscape, steelmaking then was at the heart of American industry and the soul of labor.

People moved away for careers, but not so much that you would notice. In my case, I had a fondness for the place I called home, and I had never given any serious thought to leaving Lockport, or Western New York, for that matter. Absorbed in my post high school education, however, it seemed each step I took to fulfill my career goal took me farther and farther away, at first only twenty miles to the University of Buffalo, but eventually more than 2,200 miles to Arizona in the 1970s.

From afar, I began to sense a decline in the community of my youth and the factory where my father and grandfathers had labored to support their families, and where I had spent a season as a third-generation steel-worker. Cherished memories of that time and place, where the steam hammer sounded a three-note chord that lulled me to sleep at night, became an inducement to record that intersection of personal and family labor history with a blend of steelworkers' recollections in this story, which is dedicated to them.

The Simonds Saw and Steel Company has passed into history, the buildings on Ohio Street remain fenced off, abandoned to the elements, rusting away with broken windows and missing roof tiles. Inside and out, weeds rise where workers once turned scrap metal into high-quality alloy steel. My wish to document what the place had been like earlier was a motivating factor in my decision to join my son with his camera in a journey through the abandoned plant.

This book is a tale of that journey. It weaves my story with that of many other steelworkers who spent time under the roofs that the Simonds family built. Some like me, if only briefly, were third-generation steel men who followed their father's and grandfather's footsteps across the cinder parking lot or down the Ohio Street sidewalk and through the gates of the mill. Indeed, while I began to write out of personal interest, I found that listening to steelworkers' experiences compelled me to preserve a record of labor that might otherwise be lost to the community.

This is a work of historical non-fiction with memoir thrown in: a creative journey, not a single saunter through the mill, but a composite of several visits, each taking a different route in different seasons with my participation highly imagined. The feasibility of this approach to the narrative came to me when I read Jonathon Waldman's story of photographer Eva Csuk's trip through the Bethlehem Steel Works in his book on corrosion, *Rust. The Longest War.*

We tend to understand history in terms of the written word, but meaningful history can be recorded in images. The narrative and companion photographs are meant not only to transport the reader back in time, but in adding vintage images of men at work to those of contemporary industrial decay, I hoped to draw visual attention to the impact that the abandonment of steel mills has had on workers and communities across our national landscape. Each steel community has its own story to tell, and Lockport is no exception.

This book focuses on the craft of specialty steel as Simonds made it, but will compare and contrast the manufacture of standard carbon steel in the large integrated steel mills such as Bethlehem in the context of the history behind it all. Steelmaking, particularly its technical, chemical, and metallurgical aspects, can be daunting for the general reader. I've tried to keep it simple. For those who desire more, I have included a selective bibliography, a supplementary appendix with a glossary of terms, and a list of the most common metals with their basic characteristics that Simonds used to make steel in Lockport.

Today, nearly forty years after the closure and abandonment of steel mills across the country, the subject continues to draw interest. Scarcely a week goes by without the print or broadcast media serving up a piece on the deindustrialization of America with its loss of manufacturing jobs and the plight of the blue-collar worker.

I hope this book achieves my aims, and does justice to the recollections of the men of steel. I appreciate the opportunity to share this story about the nature of work at a time and place at the heart of American industry.

—Louis A. Rosati
Mesa, Arizona

Original sketch of Simonds Saw and Steel by Joe Whalen, courtesy of John Coleman.

1
SCYTHEMAKERS

Perhaps it was my career as a pathologist that had me feeling like I was on my way to an autopsy. I thought I was prepared for this, but a nostalgic hankering to relive my steelworker days, working alongside men I'd known and respected as a kid, stirred my emotions. Nineteen fifty-nine held another halcyon summer. I was home from college, in love, playing ball with my high school pals, and making steel. Life couldn't have been sweeter. I'd read that nostalgia is a fantasy that maintains itself because it can't be fulfilled. Nevertheless, here I was, trying to recapture it more than fifty years later.

The steel industry had been a shell of itself for a good half-century. It was less competitive and less central to the US economy. That wasn't news to me, but I guess the fact that I was coming face-to-face with an indelible time and place had me wondering about a lot of things, not the least of which was coming to terms with my past.

Simonds Saw and Steel helped fashion the American dream for me and countless others. Among its employees were dozens of immigrants, such as my father and grandfathers who came to this country to work.

In their labor and frugal lifestyle they became part of a strong middle class, the likes of which we may never see again.

Somehow, I couldn't stop feeling wistful. It all led to a question I had: how did Simonds die? And because the demand for the type of steel it made never abated, there was another question: why?

It's in a pathologist's job description to seek answers in their postmortem examinations. That's part of what we do. The Simonds postmortem nagged at me as I drove over the Erie Canal on the Summit Street Bridge and saw the old mill silhouetted against the morning twilight.

"Well, here we are," I said to my son, Michael, pulling the car to a crunching stop on the cinder-packed shoulder of Ohio Street. I rolled down my window and looked out through the wild shrubs and spindly trees that filled the foreground of the fenced-in materials yard. The vegetation obscured the rail spur and lower frame of the crane trestle in the foreground of the shuttered and abandoned plant.

Michael whistled one low plaintive note while he adjusted the settings on his Nikon. I sat there downcast. *God, what happened to this place?* Finally, I got out of the car and pulled a cell phone from my pocket. Michael opened the passenger door, stepped out with his camera and reached for his backpack and tripod on the rear seat. The doors slammed in rapid succession, breaking the dawn's stillness in the gray light of the eastern sky. I leaned against the hood of the car, raised the cell phone camera and focused on a rusting corrugated-metal building with a gabled roof surmounted by a peaked superstructure. Chimneys that once belched smoke from furnaces were now silent sentinels. The chain-link fence precluded a good shot. So I stepped onto a grassy, tendril-covered sidewalk next to the fence and positioned the camera lens between the links. I zoomed in between the trees and a portion of a trestle and overhead crane to frame the buildings which, like the vegetation around it, had gone to seed. While Michael set up his equipment, I pressed the photo button. In the glow of the morning's half-light, I heard its soft click capture the starkness of what was left of a once-bustling producer of specialty steel products in Lockport, New York, the original Simonds Saw and Steel Company.

"What are you thinking about, Pop?"

My arms outstretched against the fence, my fingers enfolded in the links, I had begun the physical external examination as I flipped through pages in my memory for an old image I could not quite grasp. A sign attached to the fence to the right of my hand displayed the trefoil radioactive hazard symbol. *No trespassing.* I turned to Michael, who was adjusting his camera mount. "That was the first building they put up. It looks pretty awful now." That was not how I remembered it when I worked there. It seems like everyone I knew growing up worked in there, both of my grandfathers and their brothers, my godfather, my confirmation sponsor, most of the men in the West End. I turned to Michael. "If you had gotten a job here, you would have been the fourth generation to step under that roof."

"Yeah, interesting," he said, adjusting his lens setting. "So when did they start working here?" he asked, squinting through the viewfinder.

His question jogged my memory of my *tadone*, my grandfather, Antonio, and his time at Simonds. Both of my grandfathers had worked at Simonds, as did their brothers. "I don't know exactly, sometime after World War I. Grandpa DiPaolo started there around 1920, I think. It was about a decade or so after Simonds put up this building." My relatives told me stories about this place when I was a kid, but unfortunately I didn't pay much attention. Too busy playing in the neighborhood and fishing in the canal, I guess. It was just a summer job when I worked there in 1959, a way to make some money before going back to college for my sophomore year. I never gave the history of the place or how they made steel much thought. It was only after I started looking into the plant's history that I learned some things about the mill that they probably didn't even know. They were content to have good jobs. I don't think they were any more interested than I was about how those jobs came about. I thought about the earlier images of the plant and how different it looked now; and as Michael moved his tripod for another angle to capture the bleak landscape, I mentally wrapped text around those images.

Michael framed his view and rapidly fired off a sequence of images. "Hold it a minute, Pop," he said. "I'm going to move up the sidewalk and shoot another angle. Tell me the rest of the story when I get back. I want to know how it all started," he said as he lifted his tripod and walked off.

Our curiosities are drawn to the beginnings of things. So while I leaned back against the car and waited for Michael to finish his shots, I thought about the plant's history. I had acquired a 1921 Simonds catalog with grainy photos and descriptions of the early steelmaking process. When I first read that catalog, almost one hundred years after it was printed, I was struck by the first page, which was titled *"How to Reach Lockport."* Below it was a map drawn of Western New York with the Erie Canal, Niagara River, Lake Ontario, and the towns of Buffalo, Niagara Falls, Tonawanda, Lewiston and Lockport. The brochure read, *"the best way was to take the Lockport electric car from Court and Main Street in Buffalo and get off at the mill office in one hour and ten minutes. Ask the conductor to stop at Simonds Station . . ."*

The railway and electric car system from Buffalo that stopped at Simonds before entering Lockport's city center, and then exited the city north to Olcott Beach on Lake Ontario, was gone long before I was born at the outset of World War II; but I remember the old rails embedded in the red-brick streets when I rode my bike over them as a boy. The streets were paved over in the early fifties. Written today, the brochure would likely provide directions from the Buffalo-Niagara International Airport or the New York State Thruway. That Buffalo and Lockport were once connected by rail, and that a Simonds station had been specifically created for the plant before the train entered the city, signaled the expectation the mill would have on the local economy.

The beginning of things had to precede that catalog, but origins can be fuzzy. The family patriarch, Samuel Simonds, sailed from England in the mid-1600s before settling in Ipswich, Massachusetts. He was listed as a "gentleman" in the early records and was one of the Puritan leaders of the Massachusetts Bay Colony, becoming Deputy Governor of Massachusetts in 1673. Eight generations later, Samuel's descendent, Abel, was born in Fitchburg, Massachusetts, in 1804. It was with him that the family steel business began. During his teenage years, Abel apprenticed with a scythe and knife maker. He and his brother-in-law, John Thurston Farwell, set up shop in 1832 on the banks of the Nashua River in West Fitchburg to manufacture scythes from purchased English steel. Because Farwell was better known and experienced, the company was originally named JT Farwell and Company.

Scythes and cycles were among the most historic and simplest farming implements. Scythes became a cottage industry in Fitchburg. Literally, the factory was a cottage, and haymaking using scythes in the nineteenth century kept the little factory busy and profitable.

My father brought an old scythe home one day to cut the tall grass in a rock-strewn, empty lot beside our house on Niagara Street and Case Court where my neighborhood friends played baseball. The scythe was a relic, an instrument that had long ago been replaced by mechanized agriculture. This one had a wooden weather-beaten handle, but my father had sanded and honed the blade with a Simonds file so that the edge was relatively sharp. He showed me how to use it, but I found it difficult to swing effectively. I never did develop the knack of those early haymakers, but I nicked a lot of unseen rocks. Eventually the weeds won out and the baseball took a lot of bad hops.

In 1851 the Farwell-Simonds partnership was dissolved, but Abel continued the operation as a family business, A. Simonds and Company. Abel's eight sons worked alongside him, and the family provided nearly everything that a business and its employees required: labor, income, health and sickness benefits, paid leave, and support in retirement—ideally what every industrial worker would want today.

Abel retired in 1864 and his sons, George and Alvin, along with another investor, Benjamin Snow, formed a new business, Simonds Brothers and Company. They added machine knives, mowers and reapers to their product line. The cottage was no longer adequate to house the business, and a new facility was established in the center of town. Snow left at some point, leaving the business to be run by the Simonds brothers.

Four of Abel's eight sons left to serve with a Massachusetts regiment during the Civil War. After the war, they returned to the company, which was incorporated in 1868 as the Simonds Manufacturing Company. Five of the Simonds brothers, George, Alvin, Thomas, Edwin, and Dan, were charter members.

The brothers sold off the mower and reaper business in 1878, concentrating on saw blades and planer knives. Their products became widely known for their quality as they discovered new ways of tempering and straightening their blades. As the business grew, they opened branch offices in Chicago and San Francisco. In 1887, the company had about 200 employees; and Daniel Simonds, who began working for his father at age sixteen, became its president. He knew he could make even better saws and knives if he could find better-quality steel. The late nineteenth-century foreign steel that was bought on the open market was not always optimal. So with the idea of best quality, Dan Simonds built a mill in Chicago in 1900 and began to manufacture his own steel. With the new mill, the Chicago operation became larger than the Fitchburg parent. Simonds acquired the Canada Saw Company in 1906 and built a steel mill in Montreal, where its Canadian interests were headquartered as the Simonds Canada Saw Company. By then, the company had nine distributing branches across American and Canadian cities, and three factories in Fitchburg, Chicago and Montreal. The company had become nationally known for its fine tool steel and saw blades, having won a gold medal and Grand Prix awards at the Paris exhibition. It had thousands of employees and million-dollar assets. Growing demands of the lumber industry for Simonds saws led to expansions of the Chicago plant, where more men were employed than in Fitchburg.

By the time that the Simonds Company celebrated its seventy-fifth anniversary in 1907, Dan Simonds understood that making steel required considerable energy. Electricity was one of his greatest expenses. Finding cheaper electricity was one way to increase profitability, so he searched and found a potential steelmaking site in Western New York, a region that was emerging in the metals and steelmaking industry. Lackawanna Steel was already there (it would be purchased by Bethlehem in 1922), as were Tonawanda Iron and Wickwire Steel. Lackawanna Steel had relocated from Scranton, Pennsylvania, to the shores of Lake Erie largely through the efforts of a partnership headed by Buffalo business magnate John Albright, remembered today primarily for the Albright-Knox Art Gallery. He was a major shareholder in the Lackawanna mill and served as its director until it was purchased by Bethlehem. His syndicate acquired the Niagara, Lockport and Ontario Power Company in 1905, which had developed the hydroelectric plant and transmission facilities at the base of Niagara Falls. As president of the International Power Commission, Albright had overseen the awarding of the electrical transmission contract to George Westinghouse and General Electric, the winners of the current war between Thomas Edison's direct current and Westinghouse and Tesla's alternating current systems.

In 1910, benefiting from Albright's groundwork, Dan Simonds moved his Chicago operation to Lockport, New York, taking advantage of the hydroelectric power supplied along transmission lines from Niagara Falls to Lockport and beyond to Rochester and Syracuse. The availability of seventy wooded acres that included two farms, easy access to shipping to eastern and western markets on the New York Central and Erie Railroads, and the proximity to the expanding Erie Canal were other factors that caused Lockport to win out over several other cities that wanted the mill. In the early twentieth century, the city of 25,000 had become a good labor market with the steady arrival of European immigrants. Simonds became one of the largest mill sites in Western New York.

The harnessing of electricity and the internal combustion engine toward the end of the nineteenth century heralded one of the great rises in human productivity in the first decades of the twentieth century. It occurred during the Progressive Era when responses to "Gilded Age" problems associated with rapid industrialization included wide-ranging social and political reforms. Progressives supported scientific methods to reform and transform such diverse areas of human endeavor as education, economics, government, medicine, and industry. Abraham Flexner's 1910 report on the deplorable state of medical education in the United States and Canada serves as an example: "diploma mills," he called them. It was a watershed document that revolutionized how medical students were taught. Efficiency was an important theme; a key aspect of that efficiency was the new scientific management introduced in 1911 by Bethlehem's Fredrick Taylor ("Taylorism") that would find application in Western New York steel manufacture.

The Simonds' buildings went up amazingly fast in 1910, and when the first steel rolled out of the mills, it helped spur an economic boom in Lockport, New York. Fifty years after the first semblance of a village took shape from the impetus given to it by the completion of the Erie Canal in 1825, it began to blossom—eventually it was incorporated as a city in 1865. Several metal industries had been established there in the 1800s. Iron manufacture preceded steel when the Hall Iron Works and the Westerman Company were built in 1880 on Market and Jackson Streets, hovering around the Erie Canal and Eighteen Mile Creek. The aluminum industry that ultimately became Alcoa in Pittsburgh also got its start in Lockport. But 1910 was an especially auspicious year. Halley's Comet reappeared as Lockport not only landed Simonds, but two local citizens, Herbert Champion Harrison and Charles A. Upson, established important businesses, becoming community leaders in the process. Harrison began to manufacture a hexagonal cored, ribbon-type radiator for automobiles. His company became a division of General Motors eight years later and would become Lockport's largest industrial firm, manufacturing heating and cooling devices for all General Motors vehicles. It remains the city's largest employer. That year also saw Upson set up shop to manufacture the first easily installed 4-ply fiberboards, which became a much sought-after building material for housing, military and commercial facilities. Dan Simonds was not a local, and gets much less press in Lockport historical records; but his steel mill became a magnet in Western New York for good-paying blue-collar jobs following the erection of the new quarter-million-dollar plant on the southwestern edge of the city on the western bank of the Erie Canal. Of course, "good-paying" is a relative term that has to be adjusted for inflation. The average U.S. wage in 1910 was 22 cents an hour, the average factory worker made around $500 per year, while an engineer averaged about $5,000 per year.

That was then. In the years that followed, Simonds workers' income steadily increased as the company became a niche player in the low-carbon specialty American steel industry that was otherwise dominated by high-carbon steel producers, such as the U.S. Steel Corporation and Bethlehem. They existed on different scales in terms of product and volume. The big steel companies, the so-called "integrated" mills—behemoths that turned iron ore into structural steel (standard carbon steel), rails and beams in the millions of tons; and Simonds, a specialty mill, started with scrap metal and produced alloys (low-carbon, alloy steel), stainless and tool steel in the thousands of tons, or even in pounds for some customers. But they shared similar stories of rise and ruin that impacted not only the workers and their families, but the communities as well. The steelworkers in Lockport and Lackawanna both experienced the same emotional fallout of plant closures and abandonment, sharing in the grief, the anxiety and anger of lost livelihoods.

2
TAINTED DIRT

"**I**'m done here, Pop," Michael said, snapping me out of my reverie as he walked back down the sidewalk. The eastern horizon had begun to manifest a soft yellow glow. "Let's go in. The sun will be up and light should be good."

We knew we just couldn't walk into the place. It was a contaminated Superfund site, complete with the standard in-your-face radioactive logo deeming it strictly off-limits to the public. Michael had been in the buildings before. He was a commercial photographer with diverse interests. He was never bored by the bleak vistas of abandoned properties, but rather saw art in photographic images of urban and industrial decay. Simonds offered the perfect opportunity to photograph that genre. I enjoyed sharing that interest, and Michael was always good company. I had some trepidation, but I was feeling the need to explore the spaces where I had spent time and where my father, grandfathers and great-uncles had labored to support their families. I wanted my son to share in those memories and I thought it would give more meaning to his photographs. I was especially anxious to see what became of the sheet mill where I had worked with my father more than fifty years earlier;

and most importantly, I felt an obligation to record the Simonds story for the community on behalf of the many retired Simonds steelworkers who had been employed there. In a nutshell, we were committed.

The early morning sky was becoming brighter as the last traces of night receded on the western horizon.

"Let's go around to the north end," Michael said, as I was lost in my thoughts. "I need to get another shot from that end. Then we can come back and go in through a gap I spotted in the fence."

We got back in the car and U-turned, making a left on the narrow street named Simonds that bordered the north end of the property. In the 1940s, the property bordering the street was devoid of any structures, but now a number of small shabby warehouses and businesses lined the street on the south side.

To the north, a large field rose gradually toward the West End community where the storage towers of the flour mill were visible above a row of trees. A group of houses fronted Ohio Street to the east, and trees formed the western border of the field. When Dan Simonds erected the steel mill in 1910, he transferred 100 workers and their families from Chicago to Lockport. Among the incentives Simonds received for the move were the pledges from "moneyed men" in Lockport to erect fifty apartment-type houses on the acreage. Two-story duplexes for salaried workers were erected on New York Street less than a mile from the plant. The concept of company housing was not new. A few company towns were considered gems by urban designers; however, most of them were quite modest, and some were helter-skelter shantytowns with poor schools and municipal buildings that pressed up to the mill gates. Industrialization brought with it slum-like living conditions and pollution to some steel mill communities, but Lockport was fortunately not among them.

The Simonds family had manifested liberal tendencies in providing for its employees with housing, a benefit association and various amenities, such as the Dan Simonds Recreation Club at its Fitchburg plant. The concept found its fullest expression, however, in Bethlehem's well-planned town of Sparrows Point, Maryland. It was complete with stores, schools, churches, civic buildings, and playgrounds—virtually everything the steelworker and his family needed.

Company towns were essential to company growth and worker welfare. The Lackawanna Steel Company built a number of housing units for immigrant and migrating American families whose men came to the Western New York area seeking employment. Bethlehem continued building the community after they bought and modernized Lackawanna Steel in 1922. Ridge Road Village, Smokes Creek Village, and Bethlehem Park were three separate enclaves which included single-family and two-story homes, apartment buildings, gardens, athletic fields, even a beach on Lake Erie in what would become the city of Lackawanna, a prosperous, hard-as-steel, blue-collar town. Over the ensuing decades these, like other aging steel communities, began to show their wear.

In the 1950s, Lockport and Lackawanna High Schools were two of the eight teams in the Niagara Frontier League (our NFL). I especially remember the crowds at the games in Lackawanna. Some of the girls seemed as tough as the guys with their "DA" hairstyles and peg pants. We didn't stray far from the stands or our bus before or after the basketball and football games. The cover of our high school yearbook displayed the school logo; one of Lackawanna's was embossed with the drawing of a blast furnace. It was a tough school, in a close-knit community; kind of school you might expect to find near a steel mill, where many of its graduates found good jobs after graduation.

In providing what they considered a better life for their workers, steel company leaders defined the terms of residence, including which ethnic group could live where. They expected hard work and they expected loyalty, especially in rejecting the appeals of union organizers. This strategy worked until labor unrest began in earnest during the 1930s. In Lockport, no evidence of the fifty cottages built for the workers remained in the 1940s, when my friends and I glided our Radio Flyer wagons down the gentle slope on Ohio Street ("Simonds Street Hill"). However, the duplexes on New York Street are still there, now occupied by citizens who have no history or affiliation with Simonds.

There had been a baseball diamond on the north side of the plant just beyond the fence, where a retired steelworker told me he played ball as a kid, but that was gone, too. What I remembered was the Simonds softball team that participated in the Lockport Industrial League in the 1950s. They played their home games nearby at Grossi Park on West Avenue. Simonds workers had participated on various sporting teams since the early 1900s. Steel mill executives, including the Simonds owners, encouraged recreational activity. Baseball, softball and bowling, in later years, was especially popular among steelworkers in the various mill towns. The Lockport team even traveled to Fitchburg to play in the early days. At home they played local teams like Harrison's at their home field by the mill or at Farley Park, one of the other recreational parks in Lockport. They are depicted in a 1933 photo of an August NRA (National Recovery Act) parade in Lockport, which shows the Simonds team in all-white outfits with black lettering, posed in front of their bunting-covered truck.

■   ■   ■

I drove over the rail spur crossing that extended one mile south to the plant from the New York Central line that ran east-west from Niagara Falls to Rochester, and rolled to a stop near the fence. Through the windshield I spotted three dilapidated, paneled areas where shipping docks had once bustled with activity. Nearly all of the finished steel Simonds manufactured was shipped by rail in boxcars that were loaded from those docks. We got out, and Michael set up to take the south-facing photos in which the gabled roof building with its distinctive cupola dominated the skyline. Leaving the car where we had parked it, we walked back to Ohio Street down the sidewalk along the fence with the *No Trespassing* signs. We found the gap Michael had spotted earlier, and we made our way through it into a field filled with weeds and saplings. A hole in a second fence close to the buildings gave us access to a courtyard filled with more weeds. Michael shouldered his camera pack and held his tripod close as he curled himself in. I listened for a moment and then ducked in behind him, and we picked our way forward and worked our way to the north end of the building. Before us stood an open door flanked on either side by a tall sapling with birch-like bark and bright green leaves. Large square windows on both sides of the doorway contained squares of broken glass panes. The corrugated gray steel facing, streaked with orange and yellow-brown rust, had a pitched roof and peaked superstructure offering another photo opportunity. Eschewing the opportunity to enter at that point, we worked our way south between the building with the gabled roof on our left and the one to our right with a flat, slanted roof. We passed doors at intervals in the corrugated-metal walls that bore the radioactive trefoil and came to a covered walkway between the two buildings that had accumulated the dried fallen leaves of the previous autumn. Michael captured the image of the walkway and then he stepped through the entrance of the building on the right, the leaves crunching underfoot.

I paused and thought, *do I really want to step on radioactive ground?* Then I realized that I had spent one summer walking on that ground. There was less radiation now, although the difference was miniscule (U-238's half-life is 4.5 billion years; U-235's is 700 million years).

I stood there having a pathologist's moment, considering the preparation to begin an autopsy. I had contemplated the plant's origin and development, and realized its more current history needed updating along with a review of its systems; the time had come for an internal examination. There was no need to make a primary incision; the place was wide open, so I followed Michael in. The long building had a vaulted interior, but the floor was dirt-covered at that point, muffling our footsteps.

"Pop, do you remember this?" Michael asked, his voice echoing in the empty space. The air had that characteristic inorganic scent of abandonment. Michael had emailed me photographs with questions regarding what the various interior images represented from his earlier visit, but it had been decades since I had worked there, and I had trouble orienting my memory to the images. The plant had essentially been gutted; none of the large machines or equipment remained. It was like peering into a body in the morgue after the organs had been removed from the thoracic and abdominal cavities. I recognized little of the interior walls around me marked by a polychrome of reddish-orange and brown scale. Atmospheric oxygen and moisture had, over the years of the plant's abandonment, accelerated the corrosion we were witnessing. The rust seemed to become more intense with each Lockport visit. I thought, in time, it would rust away completely, disintegrating into nothingness.

I believed we might be standing near the midway point of what the Army Corps of Engineers (ACE), who had mapped the radioactive site, labeled as Building Number Two. It was the second building Simonds built parallel to and just west of where we had pulled up on Ohio Street. I pulled out a folded sheet from my jacket pocket that Burt Malcolm, a retired steelworker, the second of a three-generation steelworker family, had sent me. He had sketched the layout of the buildings and departments on the back of a '60s-era aerial photograph of the plant as he remembered it fifty years earlier. Burt had begun work at Simonds, joining his father there as a seventeen-year-old kid, after high school. Burt held various jobs, including in the sheet mill and its finishing department where he operated the shears. Later he became one of the electron-beam welders under the aegis of the metallurgy lab. Other assignments followed, including safety coordinator and positions at Guterl and Allegheny Ludlum, where he retired as a supervisor in the re-melting operation. Burt Malcolm, and John Coleman, another retired worker who shared many documents and stories with me, had had two of the longest careers at Simonds of all the former steelworkers I interviewed.

The original Simonds buildings in 1910 were seven in number. The first six included a gas house, melting shop (furnace building); rolling mill with the cogging and circular saw plate mill, each with annealing furnaces and shears; finishing mill with the bandsaw, hand saw and the cross cut mills each with annealing furnaces and shears; a transformer and power building for its all-electric-powered mill. The seventh building, an office and laboratory, erected outside of the general factory enclosure. Between 1913 and the 1960s, sixteen expansions and additions to the original group of structures added 520,000 square feet on 110 acres of land and altered department placements and operations. Burt's diagram reflected the additions, but the changes in the layout puzzled me.

I relied on Burt's memory as I tried to make sense of his diagram. "I don't know exactly where we are, Michael," I said. "It's been so long since I was here." I held up the diagram as I looked around the space. As far as I could tell we were in the general area of the 10- and 16-inch bar mills, the band mill, block-out shears, and cogging mill. The bar mills were not a part of the 1910 operation, which then had 250 employees; they were added in 1916. A walled-in space to our left I remembered as the site of the plant cafeteria. If I recalled correctly, it was located across from the band mill. There was a solid wall to my right cutting through the building. I thought it might be the wall I had been told about, the one the new Allegheny owners had put up after they had taken possession of the plant following the bankruptcy auction and had learned about the radioactive contamination. The wall separated the contaminated bar mills from the finishing department which they retained for future use.

I followed Michael, who was looking for a vantage point to set up his tripod. We avoided the littered areas with detritus, barrels, beams and boards. Rusting wall panels, in various hues of honey-gold to raw sienna and coffee brown beneath panels of windows, ran the length of the space. The air was clear and, apart from the soft scrape of the soles of our shoes on the floor, it was deathly quiet, not quite like I remembered the pounding roar in the summer of '59.

■   ■   ■

Back then, shafts of sunlight from clerestory windows pierced dust-laden air that foot traffic had suspended everywhere. It wasn't just the air—a thin layer of soot and dust covered nearly every surface and object from the floor to the rafters. Outdoors, black smoke rose from the furnace chimneys. Smoke curled and drifted with the breeze, prompting neighborhood women to decide whether or not to hang out the wash. Black grainy particles were present, even where men sat to eat their lunches. "It was a very dirty place," John Coleman, the ninety-year-old retired steelworker recalled. "There wasn't any dirt like Simonds dirt," he added.

It was more than dirty. It was radioactive. Modern steelmaking brought with it new Industrial-Age perils, accidents and injuries sometimes so spectacular and severe as to defy description, such as the human carnage of a blast furnace explosion. Other illnesses could be subtle and slower to develop. What neither John nor any of the other steelworkers knew then was that they had been inhaling radioactive dust on a daily basis for years.

In 1939, President Roosevelt received a letter from Albert Einstein informing him of the work of Enrico Fermi and Leo Szilard with a new element, uranium, a powerful source of energy ($E=mc^2$) that could potentially be weaponized. Concerned, Roosevelt formed a commission that became the Advisory Commission on Uranium. In 1942, the commission was integrated into the Manhattan Engineering District (MED, aka the Manhattan Project). Simonds began its involvement with the MED by rolling iron-boron hardened steel rods for the Hanford B Reactor in Washington State, where plutonium was made for the bombs that were dropped on Japan in 1945. Boron is a neutron-absorbing element that controls the fission rate of uranium and plutonium. The company took considerable pride in the work they did for the Manhattan Project.

After WWII, when all steelmakers faced a reduced demand for steel, Simonds continued to fill the needs of the new Atomic Age with its specialty alloys while some of the biggest integrated mills, such as Bethlehem, eschewed them, betting their future instead on standard carbon steel.

The Atomic Energy Commission (AEC) replaced the MED in 1946, and Simonds continued its relationship with the new government agency. In March 1948, Simonds began to roll uranium and thorium rods for nuclear reactors. Enriched uranium billets arrived at the railroad siding of the plant accompanied by government men, often under cover of darkness to maintain secrecy. Simonds would eventually double its security staff as the contract work continued.

The billets were offloaded from the rail cars and craned in to the plant. In the mill, workers loaded the estimated 200-pound, six-inch diameter, one-foot long billets into reheating furnaces using only leather gloves. The hot uranium and thorium billets were manually removed from the furnaces by hand with tongs, then subjected to grinding, before finally being rolled into 1.5-inch rods. All of the workers handling the uranium from the furnaces to the mills had radiation exposures. Burt Malcolm informed me that his father had told him that government men provided the rolling crew with white coveralls, which they collected and bagged at the end of each shift.

As the hot uranium rods passed through the steel rollers, the burnt oxidized crust sloughed off, and radioactive particulates fell onto the floor and through cracks between the floor plates into pits below the rollers. At the end of the work shift, government men wearing contamination suits with respirators swept up debris on the floor and dumped it into steel drums which were sent for reprocessing while the rods were sent to Hanford. Any of the radioactive dust that was not breathed in during the work or escaped the cleanup was re-suspended by foot traffic, and along with all of the other Simonds atmospheric particulates, it was available for inhalation.

On March 1, 1954, while steelworkers at Simonds were rolling uranium billets, a 15-megaton thermonuclear device, the hydrogen bomb, was detonated on Bikini Atoll. A giant fireball four miles wide evolved into a mushroom cloud that reached a height of 13,000 feet and a diameter of sixty-two miles in ten minutes. Seven thousand square miles of the Pacific, including a number of small atolls and ships at sea, were contaminated by radioactive fallout. The cloud drifted east, and radioactive isotopes such as Strontium-90 descended into the atmosphere and soil of the United States, eventually finding their way into the teeth and bones of '50s-era children. We now live in an environment with a higher level of radioactivity as a result of aboveground atomic weapons tests that went belowground only in the 1960s. The Bravo Test, as it was referred to in official government documents, and the Simonds exposures were both government-associated radioactive incidents, distinctions with little differences apart from scale. Government agencies considered the radioactivity from both incidents to be "essentially insignificant."

As was the case for many types of hazardous workplace exposures, the seemingly innocuous radioactivity at Simonds would become manifest many years later. Evidence of other radioactive and toxic chemical deposits around Niagara County would also materialize.

3
TOXIC PLACES

M ost steelworkers were heavy smokers, so uranium and thorium were simply other noxious toxicants in the mix of inhalants. The general public and workers had little information about the carcinogenic properties of uranium, thorium or cigarette smoke. We now realize that cancer is the culmination of mutations in one or more genes. These mutations can be acquired in several ways, including environmental insults from chemicals in tobacco smoke and radioactive elements, agents that attack DNA and change the chemical structure of the genes.

Although most denied it, some accounts suggest the workers were given film badges, but what became of the readings is not clear. For sure, the workers were not informed. In addition to the mill-side exposures, radioactive oxides were also deposited on the furnace's firebrick linings, which exposed workers whose jobs it was to reline, clean and maintain those surfaces, as well as the mill pits. My father was one of those men. During the '40s and '50s he worked in the sheet mill department during the week, but on weekends he and a coworker had a secondary job cleaning the bar mill pits to supplement their income.

"It wasn't an easy job," Burt Malcolm, said. Mike O'Donnell, another retired worker, confirmed Burt's opinion. He said, "They had to put on overalls, heavy boots and gloves and go into the pits with shovels and buckets to clean them out."

The men closest to the action each day had the heaviest exposures, ten times the federal safety standards. Many suffered no ill effects from the radioactive exposures, just as not every smoker gets cancer or heart disease, but many did; Burt Malcolm's father, Lewis, was one of the latter. He started work at Simonds when he was eighteen in the late 1930s. A 1941 group photo of the 16-inch bar mill department shows him standing tall in the back row along with fellow worker and future bar mill foreman, Harold Kinsler. Drafted into the Army during World War II, he served in Germany as a bazooka man. At six-foot-two, 170 pounds when he entered the service, he returned nearly fifty pounds lighter. At Simonds, he soon regained his weight and evolved into a strapping 235 pounds. He worked the next thirty years on the 16-inch bar mill, encompassing the years when the uranium and thorium rods were rolled. He became one of the master rollers or finishers in the '50s, a position that doubled on one of the shifts as the mill's foreman. Then his health began to fail. At first it was a heart rhythm disturbance, his son Burt told me, but it came at a time when the foreman's job became more stressful.

The bar mill crew had always functioned as a team; the *esprit de corps* and support for each man a defining characteristic of the workday. Camaraderie aside, their pay envelopes were all the more full because of it. But attitudes changed as the older men opted for easier jobs in the plant. No one could blame them. Workers had a limited life on the rolling mills. It was a younger man's job. If a worker stayed too long, he could look forward to death, disability, or chronic disease such as traumatic arthritis. However, Lewis sensed a paradigm shift in the work ethic of many of newer hires that replaced the older guys. He was offered an office job, but chose to retire with full disability. He had been raised on a farm in nearby Barker on the shores of Lake Ontario and was a part-time farmer while he worked at Simonds. There were other part-time farmers like him at Simonds, good hardworking men. At age fifty-five, he returned full-time to his small fruit and vegetable farm with its roadside stand. "He loved to farm," Burt told me. "If he could have made a living doing it, he would have, but the pay was so much better at Simonds."

He farmed for nearly thirty years, but gradually his health began to deteriorate as a result of unsuspected uranium dust exposure at Simonds. Early signs were increasing fatigue and breathlessness on exertion. When the diagnosis of chronic kidney failure was made, he wondered if the dust at Simonds could have been responsible. "There was a lot of dust," Lewis said in a newspaper interview. "We thought there might be problems. They took urine samples. Sometimes they sent us to the doctor. They always assured us there was no danger." Chronic renal disease is another known hazard of uranium exposure, as is chronic exposure to heavy metals, such as cadmium, arsenic and lead.

Over time, fluid retention and diminished urine output developed, requiring weekly dialysis treatment at a Niagara Falls hospital. Looking at the photos of a youthful man on the Simonds softball team in 1951, and a vigorous middle-aged man as he sat smiling for a photo with his bar mill crew in 1958, one wouldn't have predicted the cachectic image of a man wasting away in the last year of his life, when he died of renal failure.

Between 1948 and 1956 when the AEC contract ended, 25 to 30 million pounds of uranium and 30 to 40 thousand pounds of thorium had been processed at Simonds. The AEC contract period spanned only eight years of the three-quarter century history of Simonds Saw and Steel in Lockport, but those years of radioactive steel echo to the present day in terms of morbidity, mortality and the disappointing scene of a decaying industrial site.

The atomic weapons program was largely secretive, its risks unknown to, or rarely questioned by, the workers. The AEC was aware of the risks. They knew atomic scientists were conflicted during and after the Manhattan Project, and that research studies done on victims of nuclear tests and accidents including "downwinders" from atomic detonation tests in Alamogordo, New Mexico, and Bikini Atoll revealed elevated radiation levels. Data on the survivors of the atomic bomb attacks on Hiroshima and Nagasaki compiled by the Atomic Bomb Casualty Commission and its successor, the Radiation Effects Research Council, was also available; (Safety standards for exposure to radioactive substances date to the 1920s. By the 1940s the risks of small doses of radiation to human tissue were known.) but the agency also realized that many workers would refuse to do the work if they knew the dangers. In that Cold War era, building a nuclear arsenal took precedence.

On August 4, 1978, the Lockport newspaper, *Union-Sun & Journal*, published an article titled: "No Danger at Simonds, U.S. Inspector Finds". The Department of Energy (DOE), which had replaced the AEC, sent Dr. William E. Mott, head of the environmental technology division, to survey five radioactive sites in the area, including Simonds. Mott said in comparison with the others, he was least concerned with Simonds because he felt there was no way for the workers to come in close contact with the radioactive material that was present below the floor plates of the mills.

But in 1993, the new DOE Secretary of the Department of Energy in the Clinton administration, Hazel O'Leary, opened the classified DOE files on human radiation studies and experiments, including exposures of workers in the atomic weapons programs. By then, several Simonds workers and workers at other plants with AEC contracts were sick or had died of slow radiation poisoning. Seven years later, in 2000, Simonds was thrust into the spotlight when a sensational report in *USA Today* was published on September 6.

The report, "Poisoned Workers & Poisoned Places", prepared for the newspaper by a research institute, claimed that the medical section of the AEC and Simonds executives knew the radiation risks, but that this information was not conveyed to the workers. By 1954, the article stated, the AEC had determined that levels of uranium and thorium dust far exceeded federal standards and recommended that safety measures be upgraded and proper disposal of toxic and radioactive waste be carried out. Simonds executives, however, demurred, advising the AEC that the proposed measures were too expensive to implement and implied that canceling the government contract might be more cost effective. The AEC backed off. Simonds' work was too important to lose. Minimal precautions were taken, such as the installation of ventilation systems in some of the mills and the issuance of coveralls and respirators that were rarely used, largely because the hazards of the job were not articulated in a frank and honest way to the workers. Other precautions that might have had some impact were never implemented. It was a case of abject failure at the levels of both the federal government and local Simonds officials to protect the health of the workers, and by extension, the public.

When the exposé on the secret toxic exposure at Simonds and other sites around the country was published, indicating that officials repeatedly told workers they were fine and that specialist's warnings went unheeded, some survivors' spouses and children were quite angry. They vented their frustration publicly, blaming the workplace for their husbands' and fathers' deaths and demanding plant officials "fess up" and admit to unsafe working conditions. But some former Simonds workers met the report with skepticism. Retired steelworkers, who labored on the mills in the same jobs as the affected workers, were quoted in the *Union-Sun & Journal* as suffering no health issues attributable to the Simonds AEC contract. One of them, comfortably retired in Florida, was adamant that the report was an example of "witch hunting, sensationalism, and irresponsible journalism."

Today he would label it "fake news." Another worker, in his eighties, who was known to sometimes complain about the dirty atmosphere, said, "We were all sent to the hospital for checkups. We would go to the doctor one day and the hospital the next day. They were real good checkups." Still another said, "I'm in pretty good health, especially for someone who will turn eighty-five." However, for most of them, their family members, and community residents, the report that bad policies, poor decisions, malfeasance, and neglect could have continued years after the carcinogenic effects of the atomic bomb on downwinders living on Pacific islands, communities in the American Southwest, and Japanese survivors of the Hiroshima and Nagasaki attacks were recognized, left them shaking their heads.

In late spring, 1992, my father developed respiratory symptoms, mainly increasing cough and chest discomfort. I was sitting at my microscope in my laboratory in Mesa, Arizona, when I received a phone call from my mother notifying me that my father had developed a pleural effusion, an accumulation of fluid around one of his lungs. It had been drained and tested for cancer cells and infection, but the findings were negative. I was relieved. He was a pack-a-day Camel smoker; lung cancer was at the top of my concern list. My relief was short-lived. The effusion recurred. When drained again, the chest X-ray revealed a suspicious lung shadow. His physician in Lockport arranged for a biopsy at the Roswell Cancer Institute in Buffalo, New York. Grimly, I flew home and drove him up to Buffalo for the biopsy. We went into the radiology suite together at Roswell, where the interventional radiologist did an image-guided, fine-needle aspiration biopsy. I carried the specimen to the hospital laboratory and sat with the pathologist at a double-headed microscope, where he made the diagnosis of a *small cell undifferentiated carcinoma*, a high-grade, rapidly progressive form of lung cancer.

Surgery was not an option for that type of cancer, so a course of chemotherapy and radiation was planned. My father took the news well and never expressed any fear or anger to me. At the time, we didn't connect the cancer to the years of work at Simonds. I blamed the Camels. Cigarette smoke seemed to be a constant in our environment, in our house, our automobile, even standing by his side at the card table in Little George's Bar on those occasions when my mother sent me to remind him that it was time for supper. Forty years later he was dealt another hand from a toxic deck, but he played it well. He had always been a good bluffer.

Back in Arizona one day I received a call from his oncologist. We had spoken about the treatment plan doctor-to-doctor, but now his strategy had changed, and as a courtesy, he wanted to update me. "Your father has a large aneurysm of the abdominal aorta," he told me. "We need to treat it before we start his chemotherapy. I'll schedule a consult with a cardiovascular surgeon." As a physician and pathologist, I knew a few things about the situation. First, few people with my dad's diagnosis survived more than six months, even with treatment. Second, as a chronic smoker, my father's respiratory reserve was probably not sufficient for the major surgical procedure it took at that time to repair an aneurysm; and finally, with some angst, a sudden death from a ruptured aneurysm might be preferable than a slower and more suffering form of death from cancer. My instincts about the aneurysm repair were confirmed when the surgeon called me after his evaluation and said, "I wouldn't touch your father with a ten-foot pole."

Chemotherapy began at an oncology facility in a Buffalo suburb, close to Lockport. He tolerated the powerful poisons dripping into the veins of his arms that were now less muscled than in his steelworker days. Like most patients, he had anorexia, but fortunately minimal nausea and vomiting. The radiation therapy that followed was fatiguing, but otherwise he manifested few other symptoms.

His tumor responded at first, the X-ray shadows dwindling to nothingness; but then the tumor recurred and he began a steady downhill course. He remained stoic through it all. I spent a few days leading up to Christmas Eve with him and flew home on Christmas Day, planning to return after New Year's Day the following week. There was much left unsaid. We were both men of few words, although as I got older we communicated more easily. Saying good-bye was difficult that Christmas morning as I left for the airport. He passed away quietly two days later, seven months following his initial diagnosis.

> *And then he stops, turns and looks, just as his father stops and looks,*
> *And their eyes race across the years of unspoken love,*
> *Carrying the message that neither can speak nor bear to hear.*

From: "When a Doctor's Dad is Dying," by Robert Schwab

A few years later, a class action lawsuit was filed on behalf of steelworkers who had developed cancer. My father was one of the Simonds workers who were eligible to collect up to $150,000; however, they or their survivors who wished to make claims were required to file extensive paperwork with the U.S. Department of Labor under the Energy Employee Illness Compensation Act, including medical evidence of disease linked to the eligible site. Congress passed the act, still regarded as landmark legislation, in 2000, not long after the article in *USA Today* was published. Peter Eisler's story is credited with having had an enormous impact on the revelation and the lobbying of government legislators to pass the legislation.

Several government agencies became involved in radiation dose estimates and ancillary studies. The claims process was complicated by the fact that most of the workers were smokers, a finally-realized carcinogen. The process of filing, with the back and forth mailing of documents and questionnaires, was daunting and the adjudication was lengthy. In the case of my father, my mother's perseverance paid off. The officials concluded that despite my father's pack-a-day Camel smoking history, the lung cancer that took his life was "as likely as not" related to his employment at Simonds. In fact, the probability based on dose estimates was 85.16%. In 2006, my mother received a financial settlement check from the Department of Labor. Subsequently, the Secretary of Health and Human Services, responding to a petition filed on behalf of the workers, supported by the CDC (Center for Disease Control) and NIOSH (National Institute for Occupational Safety and Health) review boards, added the Atomic Weapons Employees of Simonds Saw and Steel to the Special Exposure Cohort. This made 600,000 former atomic weapons workers eligible for financial settlements.

Some spouses and children of the workers were denied compensation because they were unable to document with medical records the condition for which they were filing a claim: cancer, chronic beryllium disease, silicosis, asbestosis, chronic heart and kidney disease. The possible cause of Lewis Malcom's kidney disease and ultimate death was not on the list, but it might have been. Uranium can cause tissue damage as a radioactive substance and as a heavy metal. Urine test results might have helped in the claim process, but those sorts of records were often poor and incomplete. For others, documenting that they had worked at Simonds was a problem because personnel records had been dumped when the successor company, Guterl Steel, went bankrupt, closing the plant. Fortunately, in most of those cases, because the workers had paid union dues, there was a record of employment at the International Steelworker's (USW) union office in Pittsburgh.

My father's friend and coworker, Anthony Ciarfella, *'Taliano*, to most of the Italians who knew him, had worked for decades on the rolling mills and other departments at Simonds. He headed up the clean-up job on weekends with my father, shoveling out the pits beneath the rolling mills. His son, Henry, was thwarted in his attempt to process a claim on behalf of his father because although he could document his father's employment, his father's medical and hospital records had been discarded.

Another steelworker's spouse, Virginia De Voe, filed a claim but was denied. Her husband, Arthur, a tall, hearty man, had suddenly become "deathly ill" one morning on the job as a grinder. He was driven home by a fellow worker and let off on his driveway. He was able to make it to the house, but when Virginia saw how ill he looked, she rushed him to the emergency room, where he was promptly sent to the intensive care unit. He died that evening in a hypertensive crisis that caused acute heart failure. An autopsy revealed a markedly enlarged adrenal gland. It was a rare tumor, a *pheochromocytoma* that lacked the features of malignancy in terms of its propensity to spread. Just about 10 percent of them are fully malignant. It's a tumor that synthesizes and releases catecholamine (epinephrine) into the bloodstream, which caused the paroxysm at work and his death later that day. But the tumor was not on the list of twenty-two cancers eligible for compensation. Although the tumor had caused the clinical condition of "malignant hypertension," the Department of Labor, despite the family physician's appeal, concluded that the tumor did not meet the strict definition of cancer, and so it was not covered by the act. I spoke with several other steelworker family members who were either denied or frustrated in their long and tortured journey to seek compensation.

The surfaces and soils in and around the original buildings where the uranium and thorium were processed remain far above radiation safety limits today. Hold-harmless clauses in the original AEC contracts, the government claims, absolves them of responsibility. The radioactive chapter of the Simonds history ends with an epilogue yet to be written. On trips to Lockport, a drive-by of Simonds has become part of my pilgrimage. Lines from a sad poem written by Lockport native, Joyce Carol Oates, came to mind on one such visit. City of Locks was her reflection on the disappointment of a downtown renewal project gone awry from her vantage point on the Big Bridge above the locks on the Erie Canal. It would have been equally apropos standing on the Summit Street Bridge, a mile west along the canal across from Simonds:

eye to eye with broken windows . . .
across the canal we wait
wait for something to become clear
but nothing happens
in these meager cities of our childhood
nothing is declared.

Simonds wasn't the only polluter or polluted site in the city or county. Many of the city manufacturers were guilty at one time or other. Eighteen Mile Creek flows through Lockport on its course from the Erie Canal to Lake Ontario. The drainage system for the creek begins in southern Niagara and adjoining Erie counties, where rivulets and brooks gather into a stream that once coursed over a depression on the Niagara Escarpment called the Mountain Ridge. It was there that engineers decided to make a deep cut through the limestone to place the flight of five locks on the original Erie Canal.

Several chemical, paper, and plastic industries, contiguous to the creek banks, dumped toxic chemical waste into the waters. One of them, Van De Mark Chemical, a manufacturer of phosgene derivatives for plastics, and Norton Laboratories, the plastic factory where my mother worked, were situated in close proximity to the creek. When I was a boy, my friends and I would hike down over the Niagara Escarpment at Outwater Park. Heading north we could reach the creek, which in those days emitted a chemical odor from its visibly tarnished water. Everyone knew about the pollution, but there was no EPA or other agency with oversight. All sorts of chemicals and synthetic pesticides were being used in the county in those days before the Clean Water Act, and they were dumped indiscriminately without regard to long-term consequences. Today, some wish for less government regulation, but everyone wants clean air and safe water. Chronic chemical environmental exposures have not been fully appreciated, nor is there public awareness that thousands of chemicals in use then and even now have never been tested for safety.

The creek and the contaminated property along its corridor are now attempting to recover from a clean-up effort. Posted on chain-link fences along the creek are signs citing the danger of toxic waste, warning passersby to keep out. Its length of fifteen-plus miles makes it one of the largest Superfund projects in the country.

Niagara Falls, only seventeen miles away, once had a vast industrial area. In the early '50s, my family traveled through that part of the city to reach the Whirlpool Rapids Bridge over the Niagara River Gorge on our way to visit relatives in Canada. I remember the strong acrid odor, worse than anything Simonds emitted, as we passed factories that included Hooker, Dow, Occidental, Union Carbide, and Carborundum. Gypsum plants and tanneries were there as well, all of them taking advantage of the lower-cost hydroelectric power. One of those companies played a particularly toxic role in Niagara County. It was the site of one of this country's worse environmental disasters.

In 1892, a local developer, William T. Love, began a seven-mile canal to link the upper and lower Niagara River gorge, bypassing the falls. (The Welland Canal through the Niagara Peninsula in Ontario, Canada, linking Lake Ontario with Lake Erie, serves that purpose today. Love envisioned building a model community with homes and parks on the canal banks; however, an economic downturn and legal issues conspired to force Love and his investors into bankruptcy when the canal was only partially dug. The abandoned ditch became a dumpsite for the city of Niagara Falls' municipal waste in the 1920s. Other city industries also used the site for the disposal of industrial waste. Hooker Chemical acquired the property in 1942 to specifically dispose of its toxic chemical waste that included a dozen known carcinogens among the 21,000 tons of waste they eventually deposited there. Other entities, like the U.S. Army, also used the site as a waste dump. Simonds' radioactive waste was likely not deposited there, as the Army had two other disposal sites for that type of waste in the county at Tonawanda and Lewiston. At any rate, multiple users continued to contribute waste to the site until 1948, when Hooker became the sole user. In 1953, Hooker stopped dumping, covered the ditch with dirt, and sold the property to the Niagara Falls Board of Education for one dollar. The school board built an elementary school on the site and a working-class community grew up around it. The rest of the story is quickly told. A series of wet winters raised the water table and caused the chemicals to leach into the basements, yards, and playgrounds of the school and the community. A variety of illness, from asthma to epilepsy and abnormally high rates of miscarriages and birth defects, followed. In 1980, the Love Canal gained national notoriety as the first declared federal Superfund site. Over a period of several years, the community was demolished after families were relocated, and the site was finally cleaned up in 2004.

The Love Canal is just one chapter in the book on the defilement of American communities. Another chapter yet to be fully written concerns a recent revelation reported in the *Union-Sun & Journal* about scores of properties around the county contaminated by a radioactive slag byproduct of the extraction of metals from their ores. The gravel-like material, also thought to be innocuous (processed by some of the Niagara County industries mentioned earlier), was used decades ago as fill for driveways, parking lots and roads of residential, public and commercial properties.

Similar to the lead contamination of water in Flint, Michigan, the Niagara County tale of toxic dirt, dust, and sickness, is a story of negligence by industrial leaders and government agencies that failed to consider the safety of workers and the public's health. The toxic legacy lies in statistical data, which suggests that rates of several types of cancer in Niagara County are among the highest in the country.

Niagara County does not stand alone in this respect. The United States is dotted with hot spots—radioactive sites that include nuclear weapons and waste facilities as well as abandoned uranium mines. The largest of these sites is at Hanford in Benton County, Washington where underground radioactive waste storage tanks sit near the banks of the Columbia River; and where a significant slice of the Department of Energy's annual budget is committed to its long term clean-up.

In Arizona's Four Corners region, hundreds of abandoned uranium mines remain a toxic legacy to the poisoned land of the Navajo where miners and their families suffered the ill effects of radiation that had escaped as radon gas, and in mine waste, and tailings that found its way into drinking water, homes, schools and playgrounds.

4
A FLETCHER'S JOB

The searing heat was what I remembered as Michael and I moved on, passing an old annealing furnace with a corroded door. The furnaces were fixtures in the cogging, band, sheet, and bar mill departments. The heat radiating from the red-hot interiors when the doors were raised added misery to the hot humid summer days in Western New York. Workers could walk away from the furnaces, but there was no escaping the humidity. The window panels, even when cranked fully open, didn't provide enough ventilation to alter the stifling atmosphere. I was told the gantry crane men experiencing the rising heat would purposely break open the higher elevation panels and clerestory windows, whose cranks had ceased to operate. In the era before Gatorade, canisters of salt tablets were placed by water fountains so that workers could maintain electrolyte balance while they labored in flannel shirts open at the neck, exposing dirty and heavily sweat-stained long johns. Overalls draped on leather boots were streaked with dirt and scorch marks. If a worker's clothes weren't sweat and dirt-stained at the end of the day, they were likely employed in inspection and shipping; even the mill foremen looked dirty.

The heavy clothing was a necessity year round. Men could work shirtless on construction jobs in the summer, but not at Simonds, where the proximity of red-hot steel could sear skin. On hot summer days with outside temperatures in the upper 80s and 90s, the temperature in the plant could approach 140 degrees. Fans offered no relief; they just moved hot air around. It may have been a "dry heat" in the furnaces, but outside the furnace doors and elsewhere in the plant, workers' shirts were soaked with sweat. One worker reminded me, "Going outside was like entering air-conditioned space."

The landscape around the plant could be particularly bleak in the winter, with the ground and buildings covered in frost or snow. The trees were barren, and the Erie Canal was an unattractive empty ditch, having been drained for annual cleanup and maintenance. Inside the plant when workers arrived at 5:00 A.M. to heat the first charge for the 7:00 shift, it could be absolutely frigid. I could only imagine how cold it must have been for the man whose job it was to come in even earlier to ignite the furnaces. Heavy clothing, so necessary, but uncomfortable in summer, often provided insufficient warmth in winter. Apart from the area around the furnaces where steel was made or reheated, mill offices, and locker rooms, the plant was unheated. Men would cozy up to hot steel or rig up warming fires in 50-gallon barrels. The normally gray corrugated-iron walls could be white with frost, inside and out. "You could scrape it off with your fingernail," one retired worker told me. Another said, "The only things moving around in the cold were rats the size of kittens and the cats that chased them." Feral cats and rats, the other living and breathing souls at Simonds, were a permanent fixture throughout the year.

The other senses were also accosted. The activation of the electric furnace electrodes at startup could sound explosive, and its operation remained noisy. Once the shift began, dissonant sounds were pervasive.

The rings of metal on metal, and the harsh stridence of the abrasive wheels of the swing grinders on steel ingots cascading showers of sparks resembled circular saws slicing hardwood in a sawmill. In the rolling mills, the loud clangs of ingots hitting the rollers and steel tongs gripping steel sheets and bars played against the whirl of gantry cranes moving along overhead trestles and the chatter of their chains and the grind of their cables lifting and lowering loads of steel. There were one or more of these cranes moving in all the buildings as well as outdoors, and the sounds of their warning bells and the crashes of sheet metal and thuds of bundled steel bars dropping to the ground added to the cacophony. You often had to shout to be heard; and always and everywhere there was the sour smell of metals, minerals, and acid that neutralized the combusted scent of tobacco smoke. In the blacksmith, carpenter, paint, welding and millwright shops there were different ranges and intensity of noise. But the sounds I remembered most, even before I worked in the plant, were the sonorous bangs of steam hammers pounding ingots. The *"peem pa poo"* of the hammer head descending on a hot ingot, rebounding and dropping twice more, a three-note sound in the night air, reached me a mile away as I sat on my grandparents' front porch on sultry summer nights. That sound, the reincarnation of an old blacksmith at his anvil, was remembered by many residents of the west and north end of town; a sound in the night that vanished when the steam hammer was replaced by a hydraulic press in the new forge shop. There was one other sound that disappeared with the phase-out of steam: the whistle each day that signaled noon throughout the west end of town.

Michael snapped a few images while I looked at the diagram of the area again. We were standing across from the band mill, which was also called "the rope mill." It was a unique, specially designed rope-driven operation, originally made by the Mesta Machine Company and by an Allis-Chalmers motor. There is a framed large color photograph of the band mill with a painted border dedicated to the "men and women of Simonds" by the son of Harry Manning. It hangs in the hall of the former Simonds office and laboratory building outside the excised area now occupied by the Allegheny Technologies Industries (ATI) office. The photo manifests strands of thick hemp rope extending from a grooved cylinder over a trough filled with amber water to an 18-foot flywheel that was half-submerged below floor level. The photo was a realistic rendering of what Michael and I were looking at now, as were the photos he had taken of the scene in winter. The motor's housing covered the gear-driven shaft connected to the brownish-orange corroded mill housings with tarnished rollers. Two rolling stands were added to the original 1910 stand.

The crew that operated that mill took steel slabs from the annealing furnace (two Swindell oil-burning furnaces were installed in 1910) and carted them to the first of three stands where they were roughed, then finished and rolled into, long thin sheets of hack- and bandsaw thickness. One of the band mill workers who had spent about ten years working there, Dave Craine, explained to me that the 10 to 12-inch-wide bands were so long that they extended out the door into the yard. The rope was needed to supplement the motor's gears, which would otherwise stall out. Dave's foreman was Charlie "Boo-Boo" Roth. It was common practice in the band mill, as it was in many other departments, to assign nicknames to employees. The band mill probably carried the practice to the extreme; every crewmember had a colorful moniker.

Roth was the top roller. He oversaw the various jobs: the furnace heater, the front and back roughers and finishers, hookups and pullbacks. He had a seasoned crew who knew what they were doing from having rolled steel so many times, so he didn't have to shout out directions. The bands emerging from the rollers were gripped with tongs by pullback crewmen who dragged the bands across a wooden floor that was periodically hosed down to keep the hot steel moving smoothly. "Sometimes it was so smoky you couldn't see anything," Dave said. The coiling crew used tongs to gather the ends of the finished band and haul it to a coiling machine, after which it was cinctured and craned away for acid treatment in the Pickle House, and eventually the cold roll department.

The rope took a great deal of stress and wear, and it would periodically need repair and replacement. That was the job of the so-called "Fletcher"; and when that became necessary, Simonds called on men working the Buffalo shipping docks, the only personnel in the area with rope-repairing experience. The old rope was removed and laid along the factory floor. It stretched for about a quarter mile, which meant that it could extend out and along Ohio Street. Repairing the rope was not a quick and easy fix. Often the Fletcher spent a couple of days at the mill splicing in a new rope before the process resumed. It was quite an operation, and almost every Simonds worker I spoke to brought up the band mill in conversation sooner or later.

Dave may not have stood out from the crowd as a steelworker, but he was an exceptional saxophone player. I first met him when he and I were saxophonists with the Eagles Marching Band in the mid-1950s. Dave was a spirited guy, and he was especially enthusiastic when it came to women ("Sex Maniac" was his Simonds nickname) and music. After graduating high school, he attended Fredonia State College on a music scholarship. He joined the United States Air Force and toured with the band, playing for a time with the Airmen of Note. After he was discharged, Dave toured the country for over a year as a saxophonist with Eddie Condon's All Stars. He would have had a career as a professional musician or music teacher, but circumstances made him a steelworker. There were other musicians at Simonds, such as drummers Frank "Skip" Rose and Jim Fogle who, like Dave, were blue-collar workers during the week, and on weekends were members of pop and jazz ensembles at local bars and clubs. They were joined by other talented musicians employed in Lockport's many shops and factories who contributed to the rich musical culture and heritage of the town.

North of where the band mill had once been situated, three separate bar mill housings with grooved rollers stood near a bank of furnaces. They all manifested signs of disuse and corrosion. It was eerie. I glanced up through a panoply of girders and braces and a suspended curving I-beam track to gaze at cracked and broken roof panels that had once kept out the rain and snow, and thought about the time when that mill was rolling steel. I was about to comment, but when I turned, Michael had exited into a tunnel with a narrow-gauge rail line that linked to another building.

"Do you recognize anything in here?" Michael asked when I emerged. He was crossing the space toward a doorway to the exterior on the east side and hadn't given me a chance to answer. Waving me on, he said, "We'll come back here, but I just remembered the building from my last trip that I didn't get a chance to shoot. I want to be sure I get it this time."

We exited into a narrow alley filled with saplings, weeds, and scattered trees and circled around the north end of one of the original buildings Simonds built in 1910. Its corrugated metal, like the other buildings, was streaked with rust. An entrance on the east side of this building brought us into a space that manifested some resemblance to the interior of the building we had just left. It was smaller, hollowed out and abandoned to the elements, completely devoid of machinery and equipment. I unfolded another diagram John Coleman had made for me and realized that we were in the original building that housed six coal-fed Swindell-Siemens gas producers. First gas, then oil, fueled the crucible furnaces in the first years of the Simonds operation. Coal storage bunkers, conveyors, and tanks for the oil had been situated next to the gas house.

I realized that we were in the shorter building we were looking at when we pulled up this morning. The Army Corps of Engineers labeled it Number One when they did the site inspection after the radioactivity issue had become known. I hadn't been in this building before, and by the time the '50s rolled around it had been converted to other uses. It had a subterranean level filled with water that Michael said was frozen over on a winter trip he had made. The ground floor had three basic zones, and one of them had a loft with banks of multi-paned windows. The floor was strewn every which way with wood and metal debris. An old bandsaw suggested that it was some sort of wood shop. A rickety set of stairs along the north wall led to the loft that contained racks of wooden forms. We speculated that these were probably molds used in the foundry. The rest of the building included a lab-like space with kiln-type ovens that fronted a brick wall with panels of windows that admitted morning light. There wasn't much else of interest to photograph so we returned to the building we had quickly passed through, Building Number Two, which was where they had originally made the steel.

I unfolded diagrams again. Building Number Two once housed the old melt shop; and, in addition to the electric furnaces, two steam hammers, a boiler room, several work and maintenance shops, a stock room and the cast magnet department.

Pale morning light diffused in from the rows of windows along the top of both side walls. Above the windows the pitched ceiling supported by rows of triangular steel trusses admitted additional light from missing roof panels. In the center, the superstructure that ran the length of the roof was partially illuminated by the morning sun penetrating the clerestories. The echoing corrugated chamber was largely devoid of machinery that once filled the departments on John's and Burt's diagrams. Barrels and scrap metal fragments, corrugated panels, small ladles and wooden forms littered the floor, along with nondescript debris that crackled underfoot. Scattered ferns emerged from the ground. The walls of brick and metal exhibited various shades of discoloration and rust in hues of brown, orange and pale yellow, but my eye was drawn especially to the panels of corrugated green-hued Plexiglas that served to window the ends and sides of the building.

"Have you ever wondered why these Simonds buildings and practically all buildings in photos of old steel mills have this gabled design?" I asked, pointing up at the elevated ceiling.

"Not really," he said adjusting the legs of the tripod.

"It was the idea of a young engineer in the late 1800s," I said.

Fredrick Woods was waiting for a train in one of the European capitals. A locomotive chugging into the long multistory steel and glass shed covering the tracks belched smoke and soot that rose to where it escaped from the clerestories. Woods had a Eureka moment. *This might work for a steel mill. It would accommodate overhead cranes and let heat and smoke escape,* he thought. When he returned to America, he incorporated the design at the Pennsylvania Steel plant he was building at Sparrows Point, Maryland. Later, other steel mills followed that blueprint.

"Oh yeah," Michael said, wrinkling his brow as he thought back to a family trip we had made thirty years earlier. "I remember those train stations in France and Italy. Not really like this, you think?"

Woods actually used trains to link the operations in his Maryland buildings. Molten metal was poured into ladles on railroad cars and transported to the Bessemer converters, after which they were emptied into ingot molds on cars. "Casting on cars," they called it. Interplant transportation by railroad was also copied by other steel producers who had come to rely heavily on railroad transportation to move raw materials and finished product not only from site to site within the property, but to and from sources and destinations around the country.

Woods' industrial architecture was a nineteenth-century idea that steel mill operators latched onto in the twentieth century as they replaced flat-roofed mills with the gabled design. Some modified the clerestory ventilation system with roof panel windows and vents. The steel skeleton of the first Simonds buildings erected in a north-south orientation along Ohio Street resembled Woods' train shed design. The structural steel frames were overlaid with heavy, galvanized corrugated-iron siding, vented near the top, with cement tiles covering the roof. The first rolling mills were set at a right angle to those buildings; but as additions to the plant were made in subsequent years, they were attached south to north following in a parallel orientation to the buildings on Ohio Street. Space between the buildings was filled in such that workers could move from one building to the other with minimal outside exposure. The newer mills had roofs with a sawtooth design, another nineteenth-century architectural innovation. The high glass windows that received shadowless north-facing light, appreciated by artists worldwide, helped workers charged with finishing, inspecting, and shipping the final product.

Looking out a broken window, I saw the weed-filled lot we had gazed on from the street. It was definitely not like that when I worked there. "A lot of stuff came in and out of there," I said pointing to the windows. I remember work crews unloading boxcars with supplies, and also the piles of scrap metal and coal that filled the lot. There was too much material, traffic, and activity for the weeds I was now looking at to grow then. That narrow rail line we just walked along was used to carry the stuff into the plant in buggies similar to those used in coal mines. The outside crane man lifted the metal and coal into the buggy buckets and the buggies were propelled inside, where one of the inside crane men lifted the material out and deposited it where it was needed.

"They made steel from scrap, so what was the coal used for?" Michael asked.

"To heat the boilers to make steam to run the cogging hammers, I think; at least that's what I was told. The city supplied some of the steam and water and there was a backup line to the Erie Canal. The pump house is still there. The intake valve on the canal bank can be seen from the other side of the canal."

I knew that after they replaced the hammer with the press, the boilers and coal were no longer needed for that purpose. There were a lot of changes in energy needs over the years. Electricity was used to melt steel. The gas used originally for the crucible furnaces and the steam used to heat parts of the plant and run the hydraulics was largely phased out. Air pressure replaced the steam boilers. Some mill furnaces remained gas-fired, while others were fueled with oil injected through nozzles, but overall electricity was the principle source of energy.

"They poured steel in here?" Michael asked, looking around the space. "It's hard to imagine that now."

"Yes it is," I said, closing my eyes trying to picture it again.

In the summer of '59, I was assigned to what they called the labor gang. New employees started out on the labor gang and were assigned where they were most needed each day. I got to see what steelworkers called a "tap and pour," the discharge of molten steel from furnaces during an orientation walk-through on my first day of work. It was less spectacular in electric furnaces than the videos I had seen of the "blow" from blast furnaces of the big integrated steel mills where rivers of molten metal poured across platforms and down runners into ladles. But what I had witnessed on that first day of work at Simonds was thrilling enough. The Simonds steel manufacturing operation could be sub-classified into two major divisions: melting and rolling, making the steel and shaping it. As the summer wore on, I had been exposed to other departments, such as finishing, inspection and shipment, but there was no question that the melting and the rolling operations were the most fascinating. I could still envision the molten, almost white-hot steel pour out of the furnace into a ladle being held by a hook and cables from an overhead crane, and the fiery glow that illuminated the space around the men who directed the operation.

Gordon Martin, Jim McCormick, Mike O'Donnell, and John "Skeeter" Spry had all worked on the electric furnaces. They began their employment at Simonds in the fifties. Like me, they started out on the labor gang and were assigned to various departments, but eventually they acquired the seniority to apply for open positions in what was called the melt shop, the electric furnace department. A 1958 department photo of one of the old melt shop shifts showed a crew of seventeen men. Just after I had left to return to college in 1959, Gordon Martin was hired on the labor gang. Through the rest of that year as well as the next two, he worked in various departments, including the 30-inch mill as a sweep off, sand blasting, and grinding. In 1961 he was assigned to the melting department where he started out as a pit man. He progressed through the departmental job ranks to become its "first helper," steel mill jargon for the furnace operator. He became the supervisor of an entirely new melt shop department a decade later. In the '80s, after the plant was closed in bankruptcy, Gordon stayed on in an interim maintenance and security position, and was eventually rehired as the supervisor of melting in the vacuum induction unit after Allegheny Ludlum bought the operation in 1984. A seasoned and knowledgeable melting manager, hewas rewarded by the company for suggestions he made which helped improve profitability. Gordon gave me an overview of the workings of the melt shop and the basics of the electric arc furnaces.

The original electric furnaces were 6-ton oven-like units with electrodes, two long cylindrical rods that protruded from the top. The furnaces rested on the factory floor. There were three of them and each had a slag pit in front, a pour spout and side doors to admit the scrap metal. The overhead magnet crane took scrap from carts that had been rolled to the mills and placed it in piles on the floor beside the left and right side of the furnace door. The first helper led the furnace crew. He was responsible for charging the furnace with scrap metal, adding and manipulating the alloys. He had an assistant, the "second helper" who took direction from him. There were others who managed the raw materials and shoveled the scrap and alloys from left- and right-handed positions through doors on the sides of the furnace wherein it was transformed to molten metal. Right-handed men shoveled right to left, while left-handed men stood on the opposite side of the furnace, shoveling left to right as smoke billowed out the doors toward them. A few ambidextrous men could shovel from either position. Jim McCormick claimed to be a switch hitter. "I could shovel from either side," he said. A ladle man, a ladle man helper, and pit men comprised the rest of the crew.

Tony Parete was a nineteen-year-old, 129-pound young man when he started at Simonds in 1955. He had immigrated from Roccamorice in the Abruzzo region of Italy, the village of my roots that had previously populated several residences in Lockport's west end and working-class jobs at Simonds. Tony arrived well after the war in the early '50s. Shortly after he began work, longtime labor gang boss Tony D'Angelo assigned him to fill in on the night shift in the melt shop, where he was given a job shoveling scrap into the furnace. The next day he pleaded with D'Angelo saying, "Please, I can't do that shovel job. Give me something else." The boss must have taken pity, for he was reassigned, and he lasted twenty-nine more years at Simonds working in various departments. A photo of his band mill crew shows a young man with an olive complexion, dark wavy hair and a sturdy frame seated at the end of the first row. "Nothing was easy at Simonds," he said. "We did everything by hand. It was hard and heavy bullwork. There were no gravy jobs. I broke my ankle three times," he recalled, still bemoaning his working days. "I had so many injuries at Simonds I lost track." The injuries eventually led him to Simonds' bar mill finishing department where my godfather, Alphonse DiPasquale, was the boss, and where work was more easily done.

The type of steel made in the electric furnace was determined by the chemical composition of the alloys which, when added to molten metal, churned up the cauldron, as red-hot molten metal rose and then settled back. When the bath was ready for the tap and pour, the furnace was tilted and the slag was raked and poured into the slag bucket in the pit. Not all of it stayed in the bucket. The men were supposed to take turns cleaning out the slag pits once a week, Adrian Sherman told me. "I was a new hire so I got a lot more turns," he said. "It was a hot and dirty job," he added, working below a furnace where metals had been melted and poured, even well after the furnace had cooled down. Those words: *hot, and dirty*, appear repeatedly in the steelworker vocabulary. *Heavy* and *hard* are others, and for a new hire, *frightful* and *dangerous* were additional terms, albeit often left unspoken so as not to appear afraid to do the job.

Slag impurities float to the top of the melt. Visually, slag had a yellowish or gray cast that contrasted with the deeper reddish color of molten steel. It was important to distinguish between them; steel was big money, slag wasn't. Two crane men were assigned to the melting department. When the slag pot was full, the overhead crane man moved in, lowering his cable and hook, seizing the slag bucket handle, carrying it away. Outside, after it was dumped and cooled, it was recycled; some of the elements were retained and the rest sold and hauled away by an enterprising man who had a contract with Simonds to recycle the slag for its residual metal content. The high-carbon steel producers like Bethlehem allowed their slag to flow away in molten rivulets into Chesapeake Bay, or in the case of the plant at Lackawanna, transported by ladle rail cars and dumped down the banks of Lake Erie.

Meanwhile the furnace operator continued to tilt the furnace, and molten steel poured into a 6-ton capacity ladle. Some men relished being around the pour of the fiery metal; others wanted no part of it and sought jobs in other departments. When the ladle was full, crane men again moved in and carried the ladle to the ingot mold area. There the ladle man manipulated a long lever which released a ceramic ball at the base of the ladle, permitting the steel to pour into a series of ingot molds coordinated by men on the floor who set up the molds, and the crane man above who moved the ladle along. That was how the operation had been done for decades in the old melt shop, the one housed in the building where we were standing. It lasted about fifty years, until the early 1960s when the operation was moved into a new building erected on the far west end of the property and the furnaces were upgraded to 15-ton units.

Other Simonds jobs echoed in my mind from the visual cues that were present around me, and the building offered up plenty of them for photos. Michael set the Nikon on his tripod and inserted his 14–28.8 mm lens. Hunched over, he peered through the viewfinder, adjusting the focus and the aperture. A sign on the rusting wall read *"Safety Zone"* and below *"Coffee and lunch breaks only."* Green ferns sprouted from the foundation, and a leafy vine crawled up the wall next to the floor littered with corrugated-metal fragments. To the right, a greenish notice board had *"Pickle Power"* imprinted, while an adjacent open metal door displayed a *"Danger-Look out for Crane"* sign.

The Pickle House was located in its own enclosed space. Pickling in the old days was in a separate annex. Pickling and chipping were the standard treatments of steel surfaces before rolling. I remembered peering into the pickling room one day through the mist, overwhelmed by the strong acrid odor that caused me to cough repeatedly. How anyone could work in there with those fumes mystified me. I learned that the pickling environment was particularly tough on the oral cavity, particularly the teeth which could turn black, and concluded that over time workers simply became inured to the environment. But not everyone settled in.

John Coleman told me he could only take it in there a couple of days. "I counted my teeth after work to see if I had the same number I went in with in the morning," he said. Few of its workers would ever forget the noxious environment of the Pickle House. It took its toll.

Pickling, an acid-treating process, refers to the acid baths used to remove oxide scale from steel after it is formed or rolled. Some steels required it, others didn't. Neutralization and rinsing were important steps in the process. Now, seven empty rusting tanks, like tombstones, remained to document the process. One of them had raised letters, *Kolene,* another read *HCL 4* and another *No. 5 Water Rinse.* A handheld apparatus operated the crane that dunked and lifted the steel from tank to tank. An overhead crane man would have had to suffer the vapors that drifted up to him, and most likely would not have survived long in his fume-filled perch. More than one crane man in the plant had long-term respiratory problems from inhaling metal and acid fumes, which had the effect of damaging the lining cells of the airways, making the lung tissues more susceptible to infections.

I had always thought that the crane men had comfy jobs high above the fray, and in some departments they did. Actually, they performed critical work coordinating their steel-loaded cables with the men on the ground who stood in dangerous territory. With loads of steel moving overhead, mistakes could be costly; it behooved everyone to keep an eye out. Skilled crane men were sought after by department foremen and appreciated by the mill hands.

The system for moving supplies and products around inside and outside the plant involved cranes, forklifts, carts on rails, and man-powered wheeled tables called buggies which carted steel to and from annealing furnaces and rolling mills. The original buildings erected in 1910 each had overhead electric cranes with 5- to 15-ton capacities installed by the Cleveland and Alliance Crane Companies. Over the ensuing years, additional cranes with greater capacities were installed with building additions. The cranes moved in north-south or east-west directions based on the orientation of the building and girder arrangement. They charged furnaces, poured and ladled steel, manipulated and moved ingots to pre- and post-treatments, and from forges and mills to annealing, shearing, inspection, and shipping departments.

Swinging loads posed a danger, hence the warning signs we saw walking around the plant. Crane men developed department-based expertise, and the skill sets were not easily transferable when someone called in sick or overtime work was available. Looping the chains on the edges of loads and precise placement were knacks that took time and talent to master. Misplaced bundles of steel could topple onto unsuspecting workers. Dave Craine was working near the bar mill one day when the overhead crane man lowered a bundle of preheated bars onto a set of rail supports. Unfortunately, the load was dropped at the edge of the support, which tilted up, causing the bars to roll off onto the lower legs of a young man who stood close by and with his back to the descending load. Dave and coworkers, just wearing leather gloves that quickly became singed through, rushed to his side to toss the bars aside; but the man suffered a severe foot and ankle injury that essentially ended his career at Simonds. It was thought he might never walk again, but after several surgeries and a long rehabilitation, he was ambulatory once again.

5
THINGS THEY LEFT BEHIND

A sense of light and color in material decomposition struck me as we traveled from place to place. Building Number Two's length offered a variety of scenes. While little of the morning light fell on the grainy rusted panels at one end, the colors were particularly striking at the opposite end, where most of the slanted rays fell on a turquoise panel on which hung gray wooden frames. The panel was backed by linear russet hues such that the wall resembled a modern art piece. I studied the scene as I listened to the clicks of Michael's camera, then waited while he climbed to an old overhead crane. From the gantry, he captured the light that fell on a dense growth of green ferns covering the floor. I couldn't recall a sliver of green anywhere inside the plant when I worked there.

The space resembled a clandestine marijuana operation. If marijuana were legal in New York State, the abandoned space would have made an excellent growing facility—although I'm sure that the radioactivity would have added to the high in ways the user wouldn't have appreciated later. It was hard to reconcile the montage with my memories of the place in the '50s, as were photographs that Michael had taken of this building in various seasons: the vigorous green growth of spring and summer, the yellowing leaves of autumn, and the snow-covered brown vegetation of winter. The human void in those photographs was haunting.

The Simonds buildings were erected in 1910 on farm acreage with dense woods-edged meadows that had to be cleared. Each passing year since the plant's abandonment witnesses the encroaching growth of a reforming woods. Left undisturbed and given enough time, it might be possible for the trees and shrubs to reclaim the site as the buildings rust away.

When Michael climbed down, I followed him into another area with rusting pipe chases, power panels, gauges, and debris scattered about the floor. It was what remained of a maintenance area. In a side room, we discovered the old paint shop; at least that was what the sign on the fading red door read. *Jack* was lettered on an open cabinet door. John Coleman told me Jack Rising had been the painter there at one time. The benches and shelves held old Maxwell House cans, one stained with dried paint and discarded junk; debris covered the floor. Light pouring in from a pair of windows fell on a large transparent yellow smiley face attached to a wall of brown tiles with peeling white paint.

Michael paused to shoot the scene while I poked around. I found the old carpenter shop with wooden forms that were used to shape molds, and then I stumbled into a room with a rusting hood above a forge with flues that exited the ceiling. The hood had *Fat Pat* scratched into the surface. Long-handled implements were strewn about the floor in front of the forge. It was Simonds' version of an old-fashioned blacksmith shop. The blacksmith was Pat Cuillo. I knew Pat's younger brother, Dan Cuillo, a tough west end guy. *"Rough and ready for anything . . ."* read the caption under his photo in the 1956 Lockport High School yearbook; and his half-brother, John "Skeeter" Spry, would eventually land in the melt shop. That edition had photos of other graduates that would find their way to Simonds. That caption under Dan's name would have been appropriate for all of the Cuillo brothers. Five boys in a working-class family provided the setting for sibling rivalries and a kind of survival toughness they all manifested.

Most of the tools for the mill were made in Pat's shop, including the tongs and other implements that everyone used. A variety of them are seen in photographs taken during the plant's active operation. Many of them stand silently today in the shuttered buildings. Pat was a charter member of the Kingsmen, a motorcycle club founded in Lockport in 1958. The organization still has several chapters in East Coast states. Several years later Pat left Simonds to join J. Barry Hemphill's and Lou Valery's startup steel company in nearby Akron, New York, in 1982. Riding his motorcycle back to Lockport after work late one evening, he unaccountably lost control of his bike and died when he struck a roadside pole. Hundreds of motorcycle club members attended his funeral. No one could recall having seen so many motorcycles in one place; there seemed to be no end to the procession to the cemetery.

There were other empty but unremarkable shop spaces that once housed electricians, pipefitters, and welders; but when we noticed a set of iron stairs, curiosity demanded that we climb them. They led to a room above the casting floor, where we found abandoned space strewn with haphazard piles of paper, including personnel records. Later, I learned the records we had seen had been moved to the castings department from the personnel office after the steel mill was permanently shuttered. Labor union records, I was later told, had suffered a similar fate. Magazine centerfolds covered the metal ceiling and walls. The former workers in castings had found a practical use for the magnets they made on the floor below.

The locker room, one of several we explored on our journey, had wooden toilet stalls with peeling paint, lockers in which old toiletry items remained on the upper shelves, and dust-covered garments and goggles still hung from hooks. Curls of peeling paint dangled from the ceiling and walls. There was plenty of graffiti, leaving the impression that not all of the bosses or union officials were loved. The air was stale but lacked the sour rankness of sweaty work clothes. We were amazed that personal items remained in place thirty years later. Among the effects were pairs of old shoes and a dirt-encrusted hard hat with an embossed label that read *Sadlocha*. I spent a little time rummaging around while Michael took some photos of the shoes and hat. Dick Sadlocha had worked with John Coleman for thirty years. They celebrated by retiring together on the same day. John's friend had passed away a few years earlier, and when I emailed John the photo of Dick Sadlocha's hat, he wrote back, "I was shocked when I saw the picture," informing me he forwarded it to Sadlocha's widow. Unexpected discoveries of the things people carried, once wore and left behind still elicit emotional responses in the lives of people they also left behind. The old foundry had become a time capsule as well as a tomb of sorts. Satisfied with our exploration of the upper space, we descended again to the casting floor. The castings department, referred to by the workers as the "cast magnet," took up a great deal of space in Coleman's diagram. It had been a small foundry, actually a "mini mill," a mill within the mill of Simonds.

John Coleman, who spent more than thirty consecutive years at the plant, with the exception of two years in the U.S. Army in the last years of the Korean War, was one of three original employees of a department that had grown to nearly forty men. Now, more than thirty years later, we took in a rusting girded superstructure supporting a series of monorails that cast shadows on a littered floor. We followed the monorails to where they entered ovens with hanging racks and blue-gray doors marked #1, #2 and #3 in orange paint. John told me the department had its own furnaces, which were much smaller than the electric arc furnaces. Castings and ingots were poured from these smaller furnaces. The cast magnet also had its own machine shop. We came upon a pile of sand on a floor coated green with algae. The cast magnet produced several types of soft magnetic alloys and permanent magnet steels, which they poured into sand molds fashioned from a sand-molding device. The mold patterns were made to customer specification at the plant with blueprints sent by the customer, or they sent the molds. The magnets, once formed and hardened, were carefully removed from the molds and finished by grinding or on lathes. John, who operated a lathe there for many years, recalled the place as "too small, crowded and dirty for the amount of work we did, but that was the steel mill." Compared to typical steel foundries, however, the cast magnet was a miniaturization, more craft-like, and a good deal cleaner. "We had lots of overtime too," John said. "Sometimes we felt like we lived there, but the money was good when you were raising a family." That was a common sentiment echoed by many other workers.

Although I was the son and grandson of steelworkers who brought home small magnets for me to play attraction and repulsion tricks, I didn't understand the basics of magnetized steel at the time. In fact, I did not know much about how the steel produced in the casting department differed from steel made in the melt shop, or in Bethlehem's nearby steel plant at Lackawanna. Before I unraveled the attributes of the various alloys that comprised Simonds' steels, I had to understand standard steelmaking and how it differed from the steel made at Simonds. That understanding involves a detour into metallurgy, and the history of steel. It's a bit complicated and tortuous, but it's important to the Simonds story, so please bear with me.

■　　■　　■

So what is steel? When John Gunter, the American journalist and sociopolitical author, asked that question in 1947, the answer was "America," according to steel historian Mark Reuter. That response was a reflection of the role American steel played during the war, but in purely physical terms it's a type of iron with a certain carbon content, actually an alloy of iron and carbon molecules. Iron accounts for 98–99 percent, and the carbon content ranges from about 0.25-1.5 percent. Raw iron, which has 4–5 percent carbon, came to be called "pig iron" because of the way molten ore passed out of the furnace into a series of troughs, resembling a pig suckling a litter of piglets. Non-carbon or low-carbon steels, the specialty steels that Simonds produced, are made by mixing iron (ferrous scrap metal) with alloys such as cobalt or chromium; but most structural steel, such as the I-beams that Bethlehem produced, is high-carbon steel refined from iron ore.

In his book, *Making Steel*, Reutter claims there is no specific defini- tion for steel. Rather, he says, the term refers to a variety of different products that have been refined from iron ore. A complex and diverse product, it varies in size and shape—such as an I-beam or a thin steel sheet—as well as in chemical composition and metallurgical properties.

Reutter's book provides the formula for steel made by Bethlehem, a basic high-carbon steelmaker: 2 tons of iron ore, plus 1.5 tons of coking coal (bituminous coal heated in an oven to drive out impurities, leav- ing a nearly pure carbon, aka "metallurgic" coal), plus 500 pounds of limestone flux (to draw out phosphorus), and 4 tons of air equal 1 ton of pig iron. The process of converting iron into steel requires removing nearly all the carbon and then adding small amounts of it back in.

The history of steelmaking is a linear chronology, an accumulation of technical discoveries and refinements, transitions and turning points. The early discoveries came about gradually and almost accidentally toward the end of the Bronze Age. Historically, copper was among the first metals smelted (the process by which the metal is extracted from ore) around 4000 BC. A millennium later people learned to mix copper with tin to form bronze, launching the Bronze Age. Ancient blacksmiths knew the raw material, iron, was in the Earth's crust, and they also knew about meteoric iron. Eventually they learned to extract iron from its ore and forge the metal into iron tools.

No one knows precisely where, how or when steel came about, but the event probably occurred around 2000 BC in Eurasia. Jared Diamond in his book *Guns, Germs and Steel* makes a point of stating that archeological dates in metallurgical development are rather arbitrary. Some sources credit the people around the Black and Caspian Seas, the Hittites of Anatolia (part of modern day Turkey), with the discovery that iron could be made stronger by smelting with charcoal to produce a type of iron that was actually harder, but not with the hardness of future steel. The heat released oxygen, leaving lumps of relatively pure iron while adding in traces of carbon and other impurities, the slag. The blacksmiths hammered this taffy-like product into various shapes on an anvil to drive out the slag. The old Western movies portrayed the blacksmith reheating the red glowing iron in hot coals and retrieving it to form a horseshoe that he quenched in water and fastened to the hoof of a horse. That metal was soft enough to be *wrought*. But it was nearly pure iron with too little carbon, less than 0.2 percent. The other form of iron, *cast*, so-called because it could be poured directly into molds and solidified, came much later when village forge fires could be made hot enough. But this was probably a millennium before steel-making furnaces were developed.

As a teenager in Italy before emigrating, my father began apprenticing with a wrought iron maker who designed and fashioned the decorative railings characteristic of Italian balconies. Wrought iron railings flanked the front steps of our home. And the cast iron I remember with great fondness was the black cast iron fry pans, the principle cookware on my mother and grandmother's stovetop. The scent of garlic sautéing in olive oil in those pans, and the hiss of freshly added tomato greeted many awakening steelworkers on Sunday mornings. That's the extent of what I knew about iron.

Cast iron was hard but brittle, with a carbon content of 2-4 percent. These seemingly small percentage differences in the carbon content (0.2-1.5 percent) of the two types of iron are highly significant. Steel is harder than wrought iron but more malleable than cast iron. In essence, cast iron, wrought iron, and steel are similar alloys that would appear to differ little in the percentage of carbon, but greatly in strength and malleability.

Metal workers in India around 400 BC discovered that when they placed wrought iron bars in ceramic (clay) pots, added carbonaceous material, sealed them, placed them in a furnace and raised the temperature with air bellows, they could infuse the right amount of carbon as the metal melted. Cooled and solidified, they had made steel that could be hammered, bent and otherwise fashioned into swords. This *crucible* process was used for years by the early steelmakers until it was replaced by other methods and lost for a time in history.

The Persians advanced steelmaking by the process of *cementation*. This involved putting iron in stone boxes between layers of powdered charcoal and heating them. After several days the iron absorbed the right amount of carbon. The cementation process, like the crucible process, was laborious, taking days to get the temperature hot enough, and then weeks to fire up and cool the metal, break it up, reheat it and hammer it for a consistent texture. Brooke Stoddard, in his handsomely illustrated book on the history of steelmaking, *Steel: From Mine to Mill, the Metal that Made America*, describes the steel swords made by this process through the centuries, from the technological refinements of the Romans to the medieval Europeans and Japanese. Famous steel swords include the Excalibur of Arthur, the Joyeuse of Charlemagne, and the Tizona of El Cid. Japanese artisans, whose medieval swords then were of matchless quality, produced work considered treasured art today.

In the eighteenth century, Benjamin Huntsman rediscovered the crucible process that was first developed in India centuries earlier, the metallurgical knowledge moving into Western Europe at about the same time it passed south into Africa. Huntsman, an English clockmaker, was motivated to make better clock springs. After years of experimenting he developed high-quality steel at Sheffield, England, which in time became the steel capital of the world. (Dan Simonds' son, Alvan, studied metallurgy in Sheffield and brought his knowledge to the Simonds Chicago operation). Huntsman discovered that, if the less-than-optimal steel-like iron were placed in clay pots (he called them "crucibles"), then were buried in coke and started on fire, the metal would melt, thus allowing the right amount of carbon to diffuse evenly through the iron. Adding a flux to remove the fine slag particles further refined the process. This more refined crucible steel could be cast and forged into cutlery, swords, carriage and clock springs. By all accounts, the working conditions for Huntsman's steelworkers to produce this higher-quality steel were hellish. A century later, Simonds Saw and Steel would begin to manufacture their saws using steel made by the crucible process, but by then the process had become much less brutal for the worker.

Steel would be the metal that would start our country on its ascent to becoming a leader in the industry. The ratification of the Constitution of the United States of America in 1789 placed the new country on solid footing with George Washington as president and Alexander Hamilton as the first Secretary of the Treasury, then the nation's largest and most important government department.

During the colonial period, the Sterling Iron Works, founded by Peter Townsend in New York, and the Baltimore Iron Works, founded by Charles Carroll in Maryland, produced and converted pig iron to steel. The iron and steel from both companies supported the colonists in the war for independence, but afterwards Hamilton looked to the crucible steelmakers of Sheffield as a model for the industry he wanted his newly formed country to develop. Hamilton was motivated to stimulate the market in order to develop American industry and free it from the British manufacturers. He saw credit, debt, and a national bank as a means to that end; and in pleading that vision to Congress, became the principle designer of the American economy. Though Hamilton's efforts were not immediately realized, the crucible process in England would lead to the development of American specialty steelmaking in Syracuse and Lockport, New York.

In 1876, The Sanderson brothers, who began producing crucible steel in Sheffield in 1776 after Huntsman died, bought the Sweet Iron Works in Syracuse, New York. Twenty-five years later, Sanderson's Syracuse steel works became one of thirteen companies that merged into the Crucible Steel Company of America, in what became known as the "great consolidation of 1900." U.S. Steel was formed a year later in response to the Crucible merger.

Years later, the Crucible Steel Company in Syracuse, now known as Crucible Industries, would play a role in negotiations between Simonds Saw and Steel Company and the workers of the newly formed United Steelworkers chapter in Lockport. Because of the similarity in work done in the plants at Syracuse and Lockport, the Crucible wage rates were used at the bargaining table in attempting to settle one of the issues in the initial union contract.

The Sanderson brothers were among many key figures in the history of American steel. They made an important impact in Central New York, as did John Albright, who was instrumental in bringing Lackawanna Steel to Western New York. They were both less well known to the American public than the late nineteenth-century capitalists, Andrew Carnegie and Charles Schwab, who both made fortunes in steel as captains of the industry that began in Pennsylvania. Carnegie, a self-educated man of many flaws, had great drive, ingenuity, and passion, which—after a decade in railroading and investing—he focused on steel. He had the vision to grasp the innovations that were occurring in steelmaking, the genius for efficiency in mass production and the sense to vertically integrate its many processes. He was also a genius for spotting talent. It was his hiring of Charles Schwab to run his Braddock, Pennsylvania, steel works, the J Edgar Thompson Steel Mill, and the man who eventually negotiated Carnegie's consolidated steel operations, Carnegie Steel Company to J.P. Morgan, the finance wizard of the day. Morgan's purchase led to the formation of U.S. Steel in 1901. Carnegie left the steel industry to become the great philanthropist he is known as today, his name associated with concert halls, libraries, museums, and universities. In 1910, when Dan Simonds was building his plant in Lockport, Andrew Carnegie founded the Carnegie Endowment for International Peace and spent much of the last years of his life promoting world peace.

Schwab went on to become president of U.S. Steel, and eventually moved on to take charge of Bethlehem Steel after he had engineered a buyout of Fredrick Woods' Pennsylvania Steel Company at Sparrows Point. Under his direction, he nurtured Bethlehem into the giant it became, setting the tone for a philosophy of steel manufacturing, labor, and management that would continue long after his death. In his book, Mark Reutter points to Schwab and his progeny as the chief architects not only of Bethlehem's rise, but also of its ruin. Similarities in the Simonds story can be found among its pages.

But well before those innovations, another was needed. Crucible steelmaking was too laborious, slow, and expensive to make in large volumes for beams and rails. And, because it was confined to the manufacture of knives, swords, razors, springs, and tools, iron remained the metal of the early Industrial Age.

What was needed was a metal stronger than cast or wrought iron that could be produced in large volume for engineered structures—buildings, bridges, and especially for the growing railroad industry. Iron rails were too soft; they did not hold up well. Derailing accidents were common as rails would break and snake, especially on curved track which had to be replaced frequently. Moreover, railroad cars were limited in the weight they could carry. Iron was also a limiting factor in the length of bridge spans, but even relatively short spans failed, and many iron-framed structures corroded and collapsed. One of the last remaining structures from the age of iron is the Eiffel Tower in Paris. Its wrought iron surface is protected from oxidation and rust by paint; otherwise it would be history.

The breakthrough came by accident in the mid-nineteenth century in the person of Henry Bessemer, a man who had become known for his invention of bronze powder for gilding and who would go on to invent more than 100 products in iron, steel, and glass.

Certain innovations stand out in the history of steelmaking, and this was one of them. Just as the Augustinian Gregory Mendel's pea plant experiments in the monastery garden, and Watson and Crick's elucidation of the structure of DNA were turning points in the history of genetics, Bessemer's process was a game-changer in the steel mill. Some people believe the story of steel begins with him.

Bessemer was a clever man, eventually knighted for his many accomplishments. Bessemer and Benjamin Huntsman were both Sheffield men who established Sheffield as a major industrial center. But in the beginning, Bessemer knew nothing about steel. Motivated to produce military ordnance economically, he studied metallurgy and began experimenting with iron. He built a big egg-shaped converter and came up with what seemed like a foolish notion to blow air into molten pig iron. What happened next could have been a disastrous explosion, but instead something quite interesting occurred, something that revolutionized steelmaking. A rather high-intensity, 3000-degree flame was produced which burned out the carbon and impurities in the iron and created hard steel—iron with different carbon content. Actually, Bessemer was not the first to think about "air boiling," but he was the one who followed through. He had the virtue of perseverance and the knack of profiting from his inventions. When an alloy of carbon, manganese, and iron ("spiegeleisen") was added, oxygen could be removed from the molten steel and it could be poured into molds, solidified as ingots, reheated, and shaped into billets, and rolled into structural steel just by adjusting the carbon content. Bessemer's process was fast, reducing steelmaking from weeks to minutes—and because it could be made and sold in tons rather than pounds much less expensively, it launched the age of steel.

Carnegie, who had been a railroad man, and would become involved in a bridge company, appreciated the potential of Bessemer's work even as he realized that railroad men and bridge builders were not entirely happy with Bessemer metal. The steel was brittle unless it was made from phosphorus-free iron ore, found in only a few places, such as Wales, Michigan's Upper Peninsula and Minnesota's Mesabi Iron Range. The problem was solved a few years later when Sidney Thomas added limestone (Thomas Basic Process) to precipitate out the phosphorus into the slag. Iron ore from anywhere in the world could then be used to make pig iron for the Bessemer converter. Cheaper, more durable steel became available to replace wrought iron all around the world.

The Bessemer competitive advantage didn't last long. The open-hearth process, formally known as the Siemans-Martin Process after its European developers, followed Bessemer's work in the 1860s. The open-hearth furnaces were relatively broad and shallow brick-lined ovens with pipes to channel coal gas and later natural gas. Gas-ignited flames swept across the bottom of the bowl (hearth) to superheat the iron and drive out impurities. Producing a "heat" was slower than the Bessemer process, but resulted in a more durable, better-quality product. It also allowed for laboratory testing of molten steel at various intervals so that steel of precise specification could be achieved. Additionally, the open hearth permitted larger batches; and unlike the blast furnaces, they didn't need to be run 24/7, altering the shift hours of steelworkers. Steel production went from several thousands to millions of tons. Soon U.S. output exceeded that of all countries around the world.

Innovation continued in Europe even though the United States was the leading steel producer in the early twentieth century. Recall that steel made in crucibles was a high-grade steel product used mainly for tools, whereas Bessemer steel made in converters was a less-expensive metal made in large quantities. The open hearth had essentially replaced the Bessemer process, but shortly after the turn of the century (1907) the electric arc furnace (EAF), invented in France by Paul Heroult, was introduced to the United States at the Holcomb Steel Company in Syracuse, New York. Holcomb was the successor of the Sanderson-Sweet merger and the Crucible Steel consolidation. Steel of higher chemical purity could be produced entirely from scrap metal by passing an electric current through the metal generated from a high-voltage electric arc between electrodes. The manufactured graphite electrode, combined with high-power electric generation and transmission, improved furnace innings and design, creating chambers of 3000 degrees Fahrenheit— more than sufficient to melt scrap. It was also much cheaper and faster to produce steel with this method compared to the blast and open hearth furnace operations. The EAF was another game-changer, especially as it coincided with the development of metal alloys and the dawning of the specialty segment of the steel industry where the Simonds family would find their niche.

6
ALLOYED ON THE ERIE CANAL

The Erie Canal flowed beyond a copse of trees across the street from Simonds. The proximity of the plant and the canal are interesting historical coincidences in steelmaking, waterway construction and immigration. I pause here to mention this because Dan Simonds saw the canal's proximity, along with railroad spurs, as transportation links for his steel to East Coast and Midwestern markets.

At the dawn of the nineteenth century, long before William Love began his dig around Niagara Falls, other dreamers looked at the Niagara frontier and envisioned a waterway that spanned the state from the Niagara to the Hudson Rivers. The vision became a remarkable building project, an engineering marvel for its day. Started in 1817 through the political efforts of its principal proponent, Dewitt Clinton, who as governor of New York State had the passion, power, and influence to bring the project to fruition in 1825, the Erie Canal spurred development along its course. It brought Rochester, Syracuse, Utica, and Schenectady to life and made Buffalo the "Queen City," the largest grain port in the world, and New York City the country's principal commercial port on the Atlantic Seaboard. The city of Lockport arose from its creation; the locks on the Niagara Escarpment gave it its name and its nickname, Lock City.

Installing the locks on the canal was one of the last major engineering challenges. It required the use of hydraulic cement, which was patented by a young Erie Canal engineer, Canvass White, who had been sent to study the canals of Europe. Making that cement spawned a whole new American industry.

The canal's early success soon revealed the limitations of its four-foot depth and forty-foot width as passing traffic often became snarled. It was enlarged in fits and starts beginning in 1836 and finally completed in 1862. Then at the dawn of the twentieth century, during the years that Dan Simonds was moving his Chicago plant to Lockport and converting from crucible to electric steel, it was excavated to its present dimensions. It might be a stretch to say the late afternoon sun cast shadows from the rising buildings on workers widening the canal coursing along the plant's east side, but it was close.

Immigrants played a large role in building the Erie Canal. Among the many ethnic groups were the Irish in the original dig and 1840s expansion, and the Italians and Polish in the 1910–1918 expansion. Labor with picks, shovels, and black powder in the expansion was greatly augmented by steam-powered equipment and compressed air—advantages not previously available to the workforce digging the original canal. Joseph Rotondo, the grandfather of future steelworker and City of Lockport Mayor, Tom Rotondo, would intercede on behalf of the steelworkers in future negotiations with steel company executives. He was a labor contractor who recruited Italian immigrant workers to a labor camp on property he had purchased at the canal site in proximity to the future Simonds plant. Eventually, many of the canal workers and their offspring would become steelworkers at Simonds.

As a boy, I was fascinated by the view of the canal from the Big Bridge spanning seventy feet of limestone above the locks, where barges hauled material along the canal and were raised and lowered through Locks 34 and 35. What I didn't know then was one of the reasons the canal lasted as a commercial enterprise as long as it did was because the less-than-optimal iron rails limited the volume of cargo that could be carried by the railroads. Bessemer's process and the Thomas innovation that followed changed that. The average life of the rail increased from two to ten years, and the weight that a rail car could carry rose from eight to seventy tons. Steel and railroad companies advanced together, buildings rose to scrape the sky, and bridges spanned greater distances. Steel permitted John Roebling, who had bridged the Niagara River, to build the first steel bridge with a deck suspended by steel wire cable over New York's East River. The Brooklyn Bridge was revolutionary.

The nail in the coffin that brought an end to the Erie Canal as a commercial enterprise and doomed the port of Buffalo was the opening of the St. Lawrence Seaway. The canal and its locks remain woven into the fabric of Lockport's life; however, the Seaway never achieved its goal of transforming the Great Lake cities into major international ports. But it brought and still brings unintended consequences, which Dan Egan chronicles in his book, *Death and Life of the Great Lakes.*

The Erie Canal Trail Way, a walking and biking trail aside the canal, affords glimpses of the shuttered Simonds buildings in the early spring before trees develop foliage. Walking along the pathway by the Summit Street Bridge, I was reminded that one of Daniel Simonds' ideas when he decided to move his plant to Lockport, shipping by canal, never materialized. The rise of railroading, thanks to Bessemer, was the principal means of distributing Simonds steel as well as the steel made in Buffalo, Lackawanna and elsewhere. Railroads drove the Industrial Age.

■   ■   ■

Simonds Saw and Steel had begun making several types of crucible tool steel primarily for its own saws at their Chicago plant in 1900. Crucible melting in Chicago had been done in high-heat-resistant clay-graphite pots placed in melting furnaces. When the metal was molten and laboratory tested, steelworkers used long-handled ladles to pour the metal from the crucibles into molds. A later innovation was a single large ladle, which held the contents of several crucibles. It was mechanically moved along a track, raised and tilted to pour metal into a series of molds. At Lockport, the crucible process was performed in three 30-pot Swindell crucible furnaces of the Siemens type, each holding about 100 pounds of metal. However, crucible melting was soon phased out in favor of the electric arc method four years after beginning operations in 1910, just a few years after the first electric arc furnace introduced in the United States had been installed at Syracuse.

Metallurgical and analytical chemistry experiments by the turn of the twentieth century had led to the development of steels that were more corrosion-, abrasion-, and deformation-resistant; steels that could hold a cutting edge at higher heat better than standard carbon steel and could be used for making tools, machines, and springs. In 1864, Robert Mushet, an Englishman, was the first to add tungsten to his steel. His product became known as "Mushet's self-hardening steel." Fredrick Taylor, the Bethlehem consultant who would later make his name as an efficiency expert, introduced a new type of tungsten-chromium steel that became known as high-speed tool steel at the 1900 World's Fair in Paris. It typically had a higher percentage of tungsten, in addition to the chromium, percentages of vanadium, molybdenum, and cobalt. After this tool steel was introduced, machinists could run their tools several times faster, enabling, among other things, the boring of big guns for land-based artillery and battleships.

The last great challenge for the original Erie Canal builders was excavating the limestone of the Niagara Escarpment at Lockport. The carbon-steel drill bits in use at the time were brittle and broke often, frustrating the workers and engineers. Taylor's high-speed tool steel would have made drilling much easier. As it was, a Buffalo blacksmith came up with a modified method to temper and harden the steel, making the drills less brittle.

Other men experimented with alloys containing varying amounts of chromium and introduced stainless steel in the first decades of the 1900s. Simonds began with tool steel and added stainless to its product line at Lockport in the 1930s.

Stainless steel is a generic term for a group of metal alloys that have a minimum of 10.5 percent chromium, the chromium accounting for its stainless properties. Unlike iron or standard carbon steel which can corrode or rust, the result of oxygen combining with iron, the thin chromium-oxide layer on the surface of the metal forms a barrier that prevents atmospheric oxygen from penetrating beneath the surface. Also, unlike standard steel or the iron of the Eiffel tower, it needs no paint or other coating to prevent corrosion. Its other major advantage over regular steel is that when scratched it can repair itself by forming a new oxide layer, but even these properties are not absolute. Stainless does not necessarily mean stain free; it can tarnish under certain conditions. Stainless steel can be made in various grades and finishes, including high-luster surfaces, especially when formulated with nickel. The other metals found in stainless provide such characteristics as tensile strength, malleability, ductility, and hardness at high temperature. Stainless holds heat better than cast iron, and as everyone with a backyard barbecue grill knows, stainless grates are much easier to clean than cast iron grates.

These stainless steel features helped the McDonald brothers launch the fast food industry. Their San Bernardino, California, drive-in restaurant, which they opened in 1940, served a variety of items, items; but brothers decided they needed greater speed for the faster-paced society after the war. In 1948, they decided to focus on their most popular item, hamburgers, and they replaced their three-foot cast iron grill with two custom-designed six-foot stainless steel grills. The rest of the story is so well known it hardly needs repeating, but I don't think it's too much of a stretch to emphasize the impact that the heat retention and corrosion resistance of stainless steel has had on dining around the world. For that matter, look no farther than the cabinets and drawers in your own kitchen.

While reviewing Simonds' stainless products among the more than 350 different alloys they made at one time or another, I found answers to my questions about magnet steel. Stainless steel may seem all the same looking at kitchen fixtures, cookware, appliances, and medical devices and equipment, but there are low-end and high-end varieties of stainless steel in three groups. Basic stainless steel has a "ferritic" crystalline structure that gives it magnetic properties. But the most common types of stainless have an "austenitic" crystalline structure with nickel added to the chromium. The addition of nickel modifies the crystalline lattice, and strengthens the oxide layer. The dilution of iron, chromium and nickel atoms with carbon and other atoms makes the metal non-magnetic. The 18/8 stainless (percentage of chromium and nickel in the British lexicon; 304 stainless in the American system) is probably the most commonly used stainless grade. It is non-magnetic in its annealed state.

Stainless steel in contact with human tissue, "surgical steels," are essentially austenitic, chromium and nickel alloys in the "300 series"; but different grades and types are used for indwelling implants versus instruments. For example, metal stents for coronary arteries are composed of chromium and cobalt, whereas orthopedic implants may have added molybdenum, titanium, and tantalum. The third type of stainless, "martensitic," is similar to ferritic, but with a higher carbon content. It is used where strength is more important than corrosion resistance, finding use in medical and dental instruments in addition to a variety of non-medical applications.

The metallurgy of tool, stainless, and magnet steel depends not only on the chemistry but temperature, time, or speed, and method of heating and cooling. Nickel steel and other austenitic types can be made magnetic. It's a matter of chemistry and structure shaped by the annealing process. Simonds Lockport plant introduced cast magnet steel for magnetos and other uses in 1934. John Coleman told me that when he worked at Simonds they cast permanent magnet steels with various percentages of chromium, cobalt, and nickel, including 50 percent iron-nickel steel and a product called Alnico (an acronym combining the first two letters of aluminum, nickel, and cobalt). Before the use of rare earth metals, Alnico made the strongest magnets. The key to permanence in magnetic steel is in what John Coleman called a "keeper," a soft iron plate placed on the magnet's poles that *keeps* the magnet steel permanently magnetized.

In the mid-1800s, the son of an English blacksmith, Michel Faraday, explored the interaction between electrical and magnetic fields, and the Scottish Physicist James Maxwell's equations defined its characteristics. The ability to alter electron flow in electric circuitry, direction and control of magnetic flux was the function of the magnets that Simonds produced. Faraday's electromagnetic induction principle created new ways for Simonds' magnet steel to be used in a variety of products, such as magnetos, electric motors, generators, tachometers, speedometers, voltmeters, and loudspeakers. At one point Simonds produced all the magnet steel used in telephones.

Just when I thought I had a vague understanding of magnet, stainless, and tool steel, I came across another steel type that Simonds made: controlled expansion (thermostatic) bi-metals and tri-metals. Thermostatic metals perform important functions in heat control products of all sorts, such as direct reading thermometers.

Thermostats have bimetal coils or strips. The two separate metals joined together convert a temperature change into mechanical displacement, bending one way when heated and the opposite way when cooled. Steel and copper, steel and brass, or in Simonds' case, a nickel and iron core, was sandwiched and edge-welded, then hot-bonded on a rolling mill. From there it was pickled and transferred to the cold rolling department where the metal was butt-welded and cold-rolled into coils. A variety of dial and digital readout thermometers with bimetal infrastructure are on the market today. Bimetals are also found in balance wheels of clocks, band- and reciprocating saws, heat engines, electric devices, and coins, among other applications. In 1972, newly installed equipment in the cold roll department created unique edgings for bimetal bandsaw blades, and in the process, Simonds became the leading producer of bimetal edge wire for the bandsaw industry.

Less use was made of the tri-metal, composed of manganese, nickel, and copper, which have a very high expansion rate. At Simonds, the three metals could not be hot-bonded and required joining on an electron beam welder, one of the most advanced welding methods of the day. In response to the market, Simonds bought the device and sent a group of men from the metallurgy laboratory to the manufacturer in Connecticut for special training. Burt Malcolm was one of the men who operated the device. "We shipped 100,000 pounds of tri-metal to our principal customer, H.H. Wilson Company, each month," Burt said. "One month's sale paid for the cost of the welder." Shipments continued for several years until the Wilson Company shut down. Simonds' electron beam welding business followed suit. The machine was sold to a California company in 1977. The sale was a harbinger of future events. Tri-metals in use today have a different chemistry and purpose.

The new steel alloys were several times more expensive than standard steel, but they were much better for special engineering applications in a variety of industries. Hundreds of specialty steel products came on the market in the first half of the twentieth century, and Simonds provided the steel for many of them. In the second half of the century, as they continued to produce regular tool and stainless steel, Simonds made even better alloys in a new melt shop marketed for more exotic applications in the expanding aeronautics and space industry. One of their high-heat-resistant alloys became an integral component of the reentry heat shield of an early space capsule.

7
MELTING ANEW

In the early 1960s, three 15-ton capacity electric furnaces were installed in a newly completed melt shop, costing in excess of 7 million dollars. They were placed well above floor level on a tilting platform, which allowed the molten steel to be poured into larger ladles below. These furnaces, similar to the earlier 6-ton versions, had a circular shape with a lower bowl lined by refractory bricks, but with a cantilevered cover and three electrodes. At Simonds, a transformed three-phase alternating electric current was sent from the anode electrodes through the scrap metal that acted as a cathode. The ebb and flow of electricity through the metal at thousands of degrees Fahrenheit was more than sufficient to melt the metal. The new melting operation consumed more electricity in an eight-hour shift than the entire manufacturing industry in Lockport. The startup of these furnaces was startling to the ear of a newcomer, and the pour of fiery molten metal from them lit up the place with an eerie glow that was exciting to watch. Simonds' methods were considered so advanced in the new melt shop that workers remember groups of steel company executives from Japan and Italy visiting the plant to learn the latest developments in specialty steelmaking. The Texas Instrument Company honored Simonds with an achievement award in 1968.

I was curious to learn how the heats were done in the new melt shop, which had been set up after I had left for college. By happenstance, an interview with a steelworker about a completely different aspect of steelmaking led me back to it.

Mike O'Donnell grew up in one of Lockport's ethnic enclaves, the west end, which had become a predominately Italian neighborhood. Some Irish families, who had preceded the Italians in Lockport and settled in the north and west end of town, remained on the west side, and Mike spent his early days on New York Street in one of the homes that Simonds had originally built for salaried workers. He started working at Simonds in 1955 as a young man in the labor pool and transitioned through a variety of mill jobs and departments. He was an experienced worker called on to fill in from time to time in the melt shop. I wanted to know about the scrap metal, because scrap and added alloys in the electric furnace was to Simonds steel what iron ore and coke was to the steel that was born in Bethlehem's blast furnaces.

"Can you tell me about the process, starting with the scrap, Mike?" I asked. "Was it that stuff they used to pile out in the yard? I remember mounds of junk beyond the fence on Ohio Street as a kid."

"The scrap you remember," Mike said, "was a mixed bag of steel parts that included all sorts of metal junk such as cans, barrels, and hubcaps that they used to make steel in the old melt shop." He told me that for the new melt shop Simonds bought better scrap that had been separated into types. It was reduced to manageable size and trucked to the plant. "It was brought inside because it was supposed to be kept dry. There were various grades of it, and it was placed in different large, wooden bins."

"The scrap metal business became much more sophisticated over the years," John Linder told me. John had worked in his family's scrap metal business, United Alloys and Steel Corporation in Buffalo, New York, before embarking on a rabbinical career. In Scottsdale, where he is Rabbi at Temple Solel, he explained the process of moving scrap, recycled steel and ferrous metals, the raw material for 60 percent of U.S.-made steel, from its primary sources to the scrap yard and then to the mill. New steel may be composed of vehicles, appliances, leftover material from steel manufacture, and products such as cans, barrels, and demolition materials. In the scrap yard the ferrous and nonferrous metals are separated into various types and grades. The metals can be composed of large or small pieces, shredded or not, and clean or dirty. Modern salvaging machines can also sort while they shred. The various metals are then sold to brokers who deal with purchasing agents at steel mills. I wondered about new steel from old steel—*could the stainless steel on a new GE refrigerator have old Whirlpool and Frigidaire stainless elements?*

The biggest cost in an electric furnace operation is the charge material. John said, "It is much cheaper to remelt as much of a particular scrap stainless alloy as possible as to use the primary metals of iron, nickel and cobalt." Gordon Martin had mentioned that proper categorization and management of the leftover metals from rolling mill operations was important. It had as much economic importance in the mill as it had in the scrap yard. One of his duties was checking the leftover scrap alloys from the rolling operation that were supposed to be placed in properly designated bins. He was continually frustrated when his handheld meter showed discrepancies day after day between the readings he got and the bin designations. After confirming his readings with samples he took to the lab, he approached the plant superintendent with the data and his wish to resign the job. The department foremen were called together and told their jobs were on the line unless the scrap metal sorting improved. Things got a little better after that.

The composition and condition of the old scrap was not always known, and contaminants could cause problems. For example, a propane tank containing residual gas, air shocks and beer kegs with residual liquid were dangerous scrap that could explode in the furnace. "In the old melt shop an explosion was like a bomb detonating in the plant."

Water, I discovered, could be a real problem in a steel mill and could cause explosions in contact with molten steel. It's not unlike what happens when molten rock (lava) comes in contact with water, creating a volcanic explosion. Gordon Martin related a story his older brother Ray had told him about an incident in the old melt shop. The slag pit had a pumping system to keep water from accumulating; however, the pump malfunctioned one day and water pooled at the bottom of the pit. Coincidentally, the furnace sprang a leak when a worker probing the metal bath with an oxygen injector accidentally damaged the bottom of the furnace, causing a stream of molten steel to enter the slag pit. Ray Martin saw the furnace glowing red, ready to explode and yelled for the workers to run. A violent explosion blew the top and hurtled debris into the air. John "Skeeter" Spry was there that day, and he was blown back on the ground along with other crew members. The melt shop filled with billowing black smoke; nothing was visible for minutes. Workers from adjacent areas came running to help, but fortunately no one was seriously injured. The newspaper incorrectly reported that injury was avoided because the workers were out to lunch. "The only person out to lunch was the newspaper reporter," Skeeter told me. The electric furnace department had other close calls. As serious as an explosion in an electric furnace could be, it was typically less catastrophic than an explosion in one of the giant blast furnaces of the integrated steel mills where a dozen or more workers on a crew could be blown to smithereens or vaporized in an instant.

Rust was another potential problem in the old electric furnaces. Rusted metal is an oxide; heavily rusted metal contains a lot of oxygen.

Gordon recalled throwing a shovel of rusted steel into the old electric furnace and luckily turning his face away as flames shot out the door. He avoided a serious burn, but the side of his face stung for a week.

"So how did the scrap get into the furnace in the new melt shop?" I asked Mike.

"An overhead crane magnet delivered metal from the bins in the materials shop and dropped it by the furnaces. Pit men loaded the scrap into large charge buckets in a particular way called the make-up. Then the scrap was lifted to the furnace by the magnet, the roof was swung off and the furnace was charged with the metal."

The cover photograph of a Simonds catalog showed the refractory lining of the furnace roof glowing red, reflecting the radiant heat from molten metal in the cauldron below, the color contrasting with the three hot yellow electrodes in the delta zone. The light, in hues of red and yellow, illuminated the furnace operator, his assistant on the furnace platform, and the surrounding shop during the tap and pour. Charging the furnace, the order in which the various metals are placed, the positioning, timing, and the addition of the primary metal alloys, was a series of critical operations. Gordon Martin provided me the basics from his years of experience, and suggested I could find the details in a series of pamphlets he handed me that were written for electric furnace supervisors and operators by the Union Carbide Corporation. Those were items I would eventually review, but at the moment I had Mike's attention, so I asked, "What happened next?"

"It's been a while so some of the details are foggy."

I knew I was asking a lot of retired steelworkers to recall events and processes they were involved in so many years earlier. Over time, facts can soften like steel in an annealing furnace, and what is remembered about a process or incident may not be what actually took place. Still, the individual oral histories were important to document, so I asked him to continue.

"After the furnace was charged the roof was swung closed and the electrodes lowered by the furnace operator. The voltage was adjusted and the melting began."

Mike went on to explain that one or more metals, such as molybdenum ("molly" was the term the steelworkers used to avoid enunciation problems) and tungsten were added from bins by men in the melt crew whose job it was to have the alloys ready to add to the melt at the right time. The additions in the old days were done by men shoveling in the material; but in the new melt shop, a motorized device introduced the material through a furnace portal. The same portal was used to sample the bath's composition to see if it met a customer's specification. In the old days, when it was time to do this, the sample was removed and manually carried to the laboratory where it was prepped and analyzed by the chemists. Later, because time was money, a vacuum tube system was installed, cutting the transit time to twenty seconds. The results of the tests were telephoned to the furnace operator who directed the addition of more alloy, if necessary, to the melt. Every melt had an assigned heat number that identified the melt, and that number followed the product through testing, quality control, finishing, and shipment.

■   ■   ■

Simonds steel was low carbon, but not low tech in the sense that its steels were high-grade alloys and highly crafted products which reflected the scientific knowledge and skill of personnel in its chemistry and metallurgy laboratories. That was something that I didn't appreciate as a worker in the labor gang. In fact, it was the specially crafted steel in its saw blades and other products that put and kept Lockport on the steelmaking map.

The chemistry and metallurgy laboratories were housed in the two-story main office building located beside the plant enclosure, separated from the mill works. It was where the craft and creativity of Simonds flourished in a business and scientific atmosphere. The partially vine-covered, red brick structure remains standing and occupied—more than one hundred years after its steel infrastructure was erected. The large, square first-floor windows with Venetian blinds, encased by brick, and the arched, rectangular windows on the second floor impart a characteristic architecture to the building that serves as an office for its current occupant, Allegheny Specialty Metals of the Allegheny Ludlum Corporation.

The first floor housed Simonds' general administrative offices, finance and accounting. Personnel, payroll, plant supervision, credit union, and first aid were later shifted to an adjacent mill office building, a wood-framed facility that no longer stands. Laboratories for chemical, physical analysis, research, and sales were on the second floor. A new metallurgical lab with square glass-block windows became an add-on extension on the first floor at the east end of the building in 1959. Later expansions were made to accommodate data processing and the installation of a computer mainframe.

The Chicago plant, in 1910, employed a dozen men and women in its administrative office and labs, and eight men in mill supervision at the time of its relocation to Lockport. Those numbers expanded several-fold in terms of office personnel and mill supervision. A Depression-era photograph from 1933 shows several dozen personnel posing in front of the office building. Some are in work clothes while others wear suits, shirts, and ties. Unfortunately, the people are not identified, but I was excited to recognize one of them, my grandfather, Luigi, from old family photographs.

By the 1950s, when Allen Potts was the general plant manager and Don Richards was the plant superintendent in charge of production, the salaried staff had grown to around 100, which included salesmen, metallurgists, chemists, technologists, and administrative assistants. These numbers were similar to a 1967 listing which numbered ninety-three salaried staff among the total workforce of 844. These numbers would increase into the early '70s before beginning a slow but steady decline.

The main office was one place; the mill office was the other, where men and women worked together amid the clean business-like arrangements of Steelcase desks, files, and office machines in close proximity to the chemistry and metallurgical laboratories. It was a place where one might catch a whiff of estrogen in the otherwise androgenic air of the Simonds property. Business was good after the war, and the main office building, like the mill, hummed on first and second shifts. The building also had an experimental department where tempering, hardening, heating and cooling experiments were carried out. Investment of capital for research and development was integral to the long-term growth and quality of steelmaking in both the specialty and standard carbon steel plants. Contemporary critics point to a diminished commitment to research by steelmakers when the industry plateaued in the seventies, citing it as one of the factors, along with increasing labor costs, foreign competition, and decreased demand which led to the decline of the industry in this country.

Two former employees were able to fill me in on operations in the chemistry and metallurgical laboratories. Stephen Lacki started at Simonds in 1961 following a career in the Marine Corps during the Korean War. He had taken part in the epic strategic withdrawal from the Chosin Reservoir in the brutally cold winter of 1951. While he was a beleaguered marine in Korea, fighting in the snow and sub-zero temperatures, his future coworkers were beginning to earn a good salary and helping to make Simonds and Lockport prosper. After Steve left the service, he worked at the Carborundum plant in Akron, New York (later the site of Niagara Specialty Metals), while attending college at Erie County Tech on the GI Bill. Simonds subsequently hired him. While working there, he continued his education at the University of Buffalo where he earned a degree in chemistry. That was the ticket that moved him from the mill to the chemistry lab, where he had a twenty-two-year career.

Dan Balgemann grew up in nearby Middleport and began his career at Simonds in 1965. He started out in sheet mill finishing. While working, he attended night classes at the University of Buffalo where he received his bachelor's degree. The degree got him a position in the metallurgical lab. He worked with Will Radeke on bi-metals and tri-metals along with other engineers on other product lines. Later, he moved into cost accounting.

After the steel mill closed, Dan got his MBA at Canisius College before moving to Texas to manage various *maquiladora* plants, factories run by U.S. companies in Mexico to take advantage of labor costs and tax regulations. The number of those companies skyrocketed after Congress passed the North American Free Trade Agreement (NAFTA). From Texas, where Dan had retired, he informed me about the various product lines he worked on, along with the vocabulary of steelmaking: forging, annealing, quenching, and tempering that had to be performed under strict conditions of time, temperature and atmosphere. These processes and requirements were unique to each type of steel Simonds made.

The most common customer order for bar mill products when Dan worked the lab and accounting departments required hot rolling and annealing; for sheet mill products, hot rolling, annealing and pickling was needed. For all products, bar and sheet steel, the metallurgy lab certified the heat numbers, hardness, tensile strength and other physical properties such as microscopic crystalline structure, stress, elasticity, and thermal expansion. The metallurgical lab at first glance had the overall look of a business office. A partition separated the metallurgists' desks from the equipment where physical tests such as the Rockwell and Brinnell tests were carried out. The physical testing complemented and essentially certified the chemistry.

In the melt shop, alloys were added according to the chemistry sheet on a particular order, but analytic control often required multiple test samples to ensure that the melt had the precise chemical composition the customer had ordered. Bill Magrum was the chief chemist and manager of the lab where Steve Lacki and others did the analytical work. Steve Lacki became the chief chemist after Magrum retired. Now retired in Lewiston, New York, Steve offered me a collection of black-and-white photos of the chemistry lab. Its distinctive arched windows formed a backdrop for the counters of an industrial "wet" chemistry lab cluttered with containers of chemicals, beakers, graduated cylinders, Bunsen burners, ovens, balances, hoods, and analyzers. The hiss of the burners and bubbling beakers symbolized chemistry, but precise control of the steelmaking operation was provided by the "instrumental" section with its spectrographs and gas analysis units. "Did you know that the first optical emission spectrograph in the specialty steel industry was installed at Simonds?" Steve asked me. That instrument, which was built by Bill Magrum, captured quantities of the metal elements on Kodak glass plates. The plates were developed in the darkroom, and read on a densitometer. Later, a direct-reading instrument, capable of displaying quantities of sixteen different elements on a series of round clock-like meters, converted the charts to percentages, quickly relaying them to the furnace operator. Each advancement in testing reduced the time of analysis, translating into cost savings—time is money in the steel mill.

Stat testing of samples sent from the melt shop through the pneumatic tube was the lab's priority. Among the several men engaged on that second floor was my next-door neighbor on Niagara Street, Joe Apolito, my Confirmation sponsor, and *Compare*. The Apolito home was on Case Court across the empty lot adjacent to my house, which we had named "Cheater's Field." It was the grass- and weed-filled field where I had spent fruitless hours whacking weeds with the old scythe.

Joe's son, Peter, became a steelworker, and eventually a crane man, after graduating high school and after a stint in the Marine Corps during the 1958 Lebanon crisis, adding to the many three-generation families at Simonds. His father Joe, grandfather Felix, uncles John and George, and cousins, including Dave Craine, had all worked there.

Joe's workstation was by the tube's terminus in the chemistry lab. His job was to prep each sample that arrived for analysis by the chemists. He had hurt his back working on the 16-inch bar mill years earlier and was reassigned to a union job in the chemistry lab. The spinal column was suited for bending to harvest wild figs and berries of the Hunter-Gatherer Era; not lifting heavy steel ingots and bars in the Industrial Era. Many steelworkers paid a price using their backs as levers, but in the lab Joe had only to gather and leverage grams, not pounds of steel.

Joe Apolito and Steve Lacki became good friends and hunting buddies. Joe always enjoyed cooking not only the game he brought home but many other items as well. One day he told Steve and the rest of his lab crew not to bring their lunches the next day because he was making pasta with a special tomato sauce he had concocted from game. The lab looked forward to the meal, and it didn't disappoint. But later, after repeated compliments, confessed that the "game" was roadkill. Laughing, Steve recalled the event. "The sauce was very tasty," he said. But no one trusted Joe's cooking after that.

Gracie Scirto, a secretary in the office building, recalled the spaghetti dinners. "Al Ferrante was another chem lab employee who liked to cook," she said.

Al Ferrante emigrated from the Abruzzo region of Italy to the United States after the war. He arrived in 1947 with his mother; his father had emigrated before the war and couldn't return to Italy. Al met Steve Lacki at North Park School. Neither could speak the other's language, but with John Angelucci acting as translator, they became good friends. Within a year or two other future Simonds workers from the same region, Joe Santini, Tony Parete, and Frank Rosati, arrived in the United States and went to work at Simonds. Their common ethnic origin fostered close friendships among them and the several other Abruzzese immigrants at the plant. Al finished high school in Lockport and began working at Harrison's while taking classes at Erie County Technical Institute. In 1951, he left to join the Navy during the Korean War. Following postings around the Pacific and U.S. Navy ports, he resumed his studies at Erie County, but opted to join his friends in the labor gang at Simonds rather than return to Harrison's. He had taken classes in chemistry, and when a position opened in the Simonds chem lab he applied for and got the job, happily giving up working with his friends on the band and bar mills to join Steve as an analyst.

Al chuckled when I reminded him of the cooking episodes in the chem lab. He told me that he didn't cook that much; rather, he would stop on his way to work and pick up spaghetti and meatballs, anise biscotti, and a myriad other Italian treats that his mother had made. "I did that sometimes during the week, but mostly on Saturdays," he told me, adding, "I taught the guys the proper technique for dunking biscotti."

"We had lots of parties and lunches," Gracie told me. In the late 1970s, she had become the office administrator and secretary to the vice president of sales and marketing. The sales office was on the second floor across from the chemistry lab, from which various pleasant aromas would drift and invitations to share meals would be made by men cooking over Bunsen burners. "I really enjoyed working there," she said. "The salesmen were a riot, very upbeat gentlemen, so were the men in the lab. Everyone got along well; the whole second floor was very friendly."

The first floor, where plant management was ensconced, was more business-like, but Dan Balgemann confirmed that it had a family-like atmosphere. Most of the employees lived in Lockport or resided in Niagara County, so everyone knew each other and their families. The flavor in the office building on both floors changed during the Guterl years as personnel changes were made. Gracie told me she became very uncomfortable at sales meetings that involved certain executives. As the 1980s began, the administrative offices were not as happy a place as they had been, especially after layoffs started.

Simonds capitalized on a distribution system for its tool and stainless steel that was fairly unique in its day. Most of the Lockport output was sent to its "inside customers," the Fitchburg and Montreal plants, for manufacture of its Simonds brand products. The remainder was devoted to the needs of "outside" customers. The Simonds product line included hundreds of different stainless and tool steel alloys, magnet steels, and controlled expansion alloys. In the early catalogs customers were guided to the purpose for which the steel was to be used. They could choose among six categories: electric alloy steel (high-speed, hot-working, shock-resistant, water-hardening, cold-worked, and special-purpose types); cutting (saws); shearing (blades and slitters); forming (press dies); hammering (chisels) or rolling (at other mills). Over the years, sales and marketing personnel assisted their increasingly sophisticated outside customers with specific choices based on newer chemical compositions and metallurgic properties.

Dan Balgemann told me that Simonds made stainless alloys that others made, such as 304 stainless, found in many kitchen and bathroom fixtures, but also alloys that no one else made. One of Simonds' 300 stainless series, High Temperature Super Alloy RA-333, was developed and patented there. Many credit the metallurgic creativity of Laurence Van Mater for that and other alloys in Simonds' successful product line. "We were lucky to have him at Simonds," more than one former employee told me. Alloy 333 was composed mainly of nickel and chromium, but also had small percentages (3 percent each) of tungsten, cobalt and molybdenum, which made it highly heat-, acid-, and shock-resistant. It was an expensive alloy for the time, but its cost was offset by its increased service life.

The nineteenth-century Russian chemist, Mendeleyev, developed a chart, known as the *Periodic Table of Chemical Elements*, a tabular arrangement of the chemicals in columns (groups) and rows (periods) based on their atomic number and relatedness. This chart lies at the heart of Simonds metal alloys. Each element is listed with its atomic number and weight. The atomic number refers to the positive charge of the atom's nucleus, i.e., the number of protons. The atomic weight is the sum of the atomic number and the number of neutrons in the nucleus. Critically important to each element and its interactions are the electrons orbiting the nucleus; however, they are not directly notated in the chart.

The table, which has been expanded over the years to contain 118 elements to date, includes all of the metals that Simonds used to make their steel, including titanium, vanadium, manganese, iron, cobalt, nickel, and copper, which are listed sequentially left to right in Period 4. Each element in the period has one more electron than the neighbor to its left. Intimate relationships are also shown vertically. In the column below chromium are molybdenum and tungsten in Periods 5 and 6, which have been placed in those positions because of the configuration of the electrons.

That chart was a mainstay on the wall next to the blackboard in my high school and college chemistry classrooms. It was my first exposure to the molecular aspects of those metals, but at the time I didn't appreciate the applied practical relevance it had for Simonds steel. For steelmakers, those metals had unique properties, and they were added to the scrap iron melts like chefs added condiments, herbs, and spices, or like bakers added flour, sugar, and shortening to make a cake.

Culinary recipes typically have a list of ingredients and a process to combine and treat them. Simonds had recipes for blending each alloy in the right amount at the right temperature and the right amount of time to make its various steels. There were standard steels with well-defined chemical and physical characteristics that a customer could choose from, just as a customer could walk into a bakery and choose a standard layer cake: white, chocolate, or yellow. But some customers wanted the recipe tweaked, a marble or strawberry cake, or had a more customized product in mind, like a strawberry-orange ricotta cake with macadamia nuts and coconut. Simonds' strength was that it could meet a customer's needs by adding condiments to the tap with its special recipes, as long as the customer was willing to buy the whole cake and not just a slice. (Niagara Specialty Metals, a later start-up in Akron, New York, succeeded in part because it was willing to make and sell a slice.) And just as the baker had a post-oven cooling routine, tested his cake with a toothpick, and finished the surface and layers with custom icing, the metallurgy lab tested the physical characteristics, and finishing departments applied surface luster or other treatments.

I intended to delve deeper into the chemical and metallurgical details of the various alloys. Many of them have a fascinating history—but the information does not make for lively reading—so I place the Periodic Table here to illustrate the relationships of the Simonds elements, relegate the details to the appendix, and return to the melting operation.

"What happened in the melt shop after the call from the chem lab?" I asked Mike.

"Well, when the chemistry and temperature was right," he said, "a bell sounded which indicated that the steel was ready for the pour."

"Did they get the melt right on the first try?" I asked.

"Yes, usually the preliminary test was good; sometimes two or three additions of an alloy and tests were needed, but that didn't happen often."

Periodic Table of the Elements

http://chemistry.about.com
©2012 Todd Helmenstine
About Chemistry

1A																	8A
1 H 1.00794	2A											3A	4A	5A	6A	7A	2 He 4.002602
3 Li 6.941	4 Be 9.012182											5 B 10.811	6 C 12.0107	7 N 14.0067	8 O 15.9994	9 F 18.9984032	10 Ne 20.1797
11 Na 22.989769	12 Mg 24.3050	3B	4B	5B	6B	7B	← 8B →			1B	2B	13 Al 26.9815386	14 Si 28.0855	15 P 30.973762	16 S 32.065	17 Cl 35.453	18 Ar 39.948
19 K 39.0983	20 Ca 40.078	21 Sc 44.955912	22 Ti 47.867	23 V 50.9415	24 Cr 51.9961	25 Mn 54.938045	26 Fe 55.845	27 Co 58.933195	28 Ni 58.6934	29 Cu 63.546	30 Zn 65.38	31 Ga 69.723	32 Ge 72.64	33 As 74.92160	34 Se 78.96	35 Br 79.904	36 Kr 83.798
37 Rb 85.4678	38 Sr 87.62	39 Y 88.90585	40 Zr 91.224	41 Nb 92.90638	42 Mo 95.96	43 Tc [98]	44 Ru 101.07	45 Rh 102.90550	46 Pd 106.42	47 Ag 107.8682	48 Cd 112.411	49 In 114.818	50 Sn 118.710	51 Sb 121.760	52 Te 127.60	53 I 126.90447	54 Xe 131.293
55 Cs 132.9054519	56 Ba 137.327	57-71 Lanthanides	72 Hf 178.49	73 Ta 180.94788	74 W 183.84	75 Re 186.207	76 Os 190.23	77 Ir 192.217	78 Pt 195.084	79 Au 196.966569	80 Hg 200.59	81 Tl 204.3833	82 Pb 207.2	83 Bi 208.98040	84 Po [209]	85 At [210]	86 Rn [222]
87 Fr [223]	88 Ra [226]	89-103 Actinides	104 Rf [267]	105 Db [268]	106 Sg [271]	107 Bh [272]	108 Hs [270]	109 Mt [276]	110 Ds [281]	111 Rg [280]	112 Cn [285]	113 Uut [284]	114 Fl [289]	115 Uup [288]	116 Lv [293]	117 Uus [294]	118 Uuo [294]

Lanthanides	57 La 138.90547	58 Ce 140.116	59 Pr 140.90765	60 Nd 144.242	61 Pm [145]	62 Sm 150.36	63 Eu 151.964	64 Gd 157.25	65 Tb 158.92535	66 Dy 162.500	67 Ho 164.93032	68 Er 167.259	69 Tm 168.93421	70 Yb 173.054	71 Lu 174.9668
Actinides	89 Ac [227]	90 Th 232.03806	91 Pa 231.03588	92 U 238.02891	93 Np [237]	94 Pu [244]	95 Am [243]	96 Cm [247]	97 Bk [247]	98 Cf [251]	99 Es [252]	100 Fm [257]	101 Md [258]	102 No [259]	103 Lr [262]

Principle Simonds Elements

I knew that the ability to test and adjust the chemistry was one of the big advantages of electric alloy steel. "You told me about how the testing was done in the old days. How did the process change?"

"Much later on the AOD was added to the electric furnace operation and that made it easy."

Mike explained that the AOD stands for argon, oxygen, and decarburization. In illustrations, it looks like a big pot that sits on a trunnion that can be rotated and tilted.

"Think about the EAF and the AOD as a combined operation," Mike said. "The AOD takes most of the carbon out of the molten steel and also removes most of the impurities."

The AOD unit was another inflection point in the specialty steel-making story. It was introduced in 1954 by the Linde Division of Union Carbide for the manufacture of stainless and other high-grade steel alloys. Stunning photographs of the operation show the electric arc furnace set at a 90-degree angle, pouring white-hot molten steel into an AOD unit tilted forward to receive the charge. Injected argon and oxygen gases stir the brew, alloys are added to recover chromium, and temperature and chemical analysis is performed periodically. When the molten metal was ready, the vessel could be heeled over to pour off the slag containing impurities such as sulfur and phosphorus into slag pots; the rest is poured into ladles. The AOD made the entire steelmaking process more efficient, more easily tested, and it significantly improved the quality of the steel; but working in the melt shop could still be hazardous. Engineering design, work procedure, and rules kept one safe to the extent that precautions were taken.

8
RISKY BUSINESS

M ike was a ladle man when he worked in the melt shop. That was
also Skeeter Spry's job much of the time. They manipulated the
ladle's stopper rod, which controlled the ceramic ball that covered the
hole on the bottom of the ladle. The overhead crane operator then lifted
the ladle to a row of molds where the molten steel was transferred in a
sequential fashion as in the old melt shop. In time, the crane operator
was replaced by a pushbutton remote-control device that one of the crew
operated. The technologic innovations in the melt shop led to better steel
and increased efficiency, but for the workers it remained a risky business.

One day when Mike was on top of the ladle adjusting the stopper rod,
he was accidently knocked off the ladle by the cable assembly controlled
by the remote-control operator. He fell about ten feet to the concrete
floor, fracturing his pelvis. He was hospitalized, spending several weeks
on his back in traction before recovering sufficiently to perform light
duty and then his job in the melt shop. His later years at Simonds were
spent as an annealing furnace operator, monitoring the heating process
and making the necessary adjustments.

Today steelworkers labor under safer conditions, a situation unlike the steel mills of the nineteenth and twentieth centuries, where danger lurked everywhere. Previous chapters have cited some of these dangers: Radioactive and heavy metal (cadmium and beryllium) exposure, explosive injury, burns, lacerations, crush injuries, fractures, electrocutions, and concussions from blows of swinging and flying objects were some of the many threats to steelworker's lives and limbs. Burns, bumps, and bruises were minor everyday problems. "People were always getting stitched up," I was told. And unlike chronic heart, lung, kidney disease, spinal problems, and arthritis that developed slowly over time, these acute traumatic risks could come at you quickly and from any direction in the form of hot steel bars through a set of rollers, a press, or a crane carrying a load of metal. Financial awards for acute industrial injuries and disability were made weekly at the Compensation Court held in Lockport's City Hall in the 1950s. Typically the disbursement totaled several thousand dollars each session.

One of the stories that a retired steelworker related, but hadn't personally witnessed, had been told to him by a Simonds worker a generation older. I thought it apocryphal when I heard it repeated by another retired Bethlehem-Lackawanna steelworker. Both incidents involved a worker on the pouring platform being accidentally knocked into a ladle of molten steel by a crane. The metal and the consumed body within was hauled out of the plant and buried in the yard because the pit crew refused to pour the contents of the ladle into molds or handle it in any way. It may be a universal steelworker legend. Simonds' employment manager, Dave DeLang, told me he heard the same story on a visit to a Pittsburgh steel mill.

Skeeter Spry's first permanent assignment in the plant after the labor gang was the melt shop. On the first day on the job, one of the senior men sent him on a fool's errand, although he didn't know it at the time.

"Get me a 'canuter valve' from the cold roll," he was told. He had no idea what the device was and failed to notice the smirk on the faces of the other men. An hour later, after a wild-goose chase through several departments and buildings for a nonexistent item, he returned frustrated and empty handed, only to be chewed out by the melt shop foreman. It was part of a hazing process for newcomers. The snipe hunt was not unique to the melt shop. The rolling mills had a number of pranksters among its mentors.

Skeeter was still relatively new on the job when the electric furnace failed to arc after new electrodes had been lifted to the top of the furnace, aligned, and screwed into their socket. The graphite electrodes take a beating, and the electric arc itself constantly gnaws away the tips with continued use; the graphite subliming from a solid to a gas. Working around the electric arc furnace carried risk. Skeeter climbed to the top of the furnace to make a manual adjustment on the screw-in device with a long metal wrench ("wishbone spacer") so that the electrodes would seat properly. He looked to the furnace operator who nodded that the power was off.

A key component of the EAF is the transformer that receives thousands of volts of electricity, generated and transmitted from the power utility. The transformer steps down that voltage to what the furnace design requires; however, considerable voltage remains in the circuit. A cardinal rule in the melt shop required the furnace operator to lock out the power anytime someone made a trip to the top.

But as Skeeter placed his foot and made contact with the wrench, he triangulated a circuit, and in his words, "was gone." He said, when I asked what that meant, "I was arced off the furnace." When he revived, he discovered that everywhere he had metal touching or close to his skin, he was burned; his toenails from his steel toe boots; the skin of his finger from his ring and his abdomen from his belt buckle were heavily blistered. It could have been worse. If he had been grounded, he would have been toast.

■　■　■

The melt shop operation did not end with the pour into molds. One more step was required before the steel left the shop. "I want to finish up with the melt shop operation, Mike. Take me back to what happened after the pour into the molds."

"A hot top was placed on top of the mold to trap the impurities that rose as the steel cooled," he said. "It also improved the solidification."

The hot top Simonds used in the old days was a cast iron device of their design containing refractory made from a dry mortar mix that resembled wheat flour. Contemporary electric furnaces and hot tops are lined with a ceramic refractory that protects the roof, sidewalls, and bottom. Ceramics are oxides, carbides or nitrides of such metals as magnesium, aluminum, zirconium, silica or graphite. Fire bricks composed of these materials are placed against a fireproof lining which protects the furnace sidewall. In the early- to mid-twentieth century at Simonds and other steel mills, asbestos was preferred because it was cheap, easy to use, readily available, and not thought to be hazardous. Heavy asbestos exposure was another hidden danger in the plant that most other workers didn't appreciate.

"Do you remember the hot top process?"

"When I worked the melt shop what we called 'hot top' came in bags, but what was in the bags I'm not sure. It looked floury. We mixed it with water and poured it into the hot top molds."

"What happened next?"

"The steel had to be slowly cooled; it took about an hour."

Although the quenching of steel might seem like watching paint dry, at the microscopic level a major transformation takes place as the grain structure of the steel is altered by the rearrangement of metal atoms. The alteration imparts steel's versatility, its softness and malleability or its hardness and resistance to corrosion. An aberrant rearrangement of those atoms under less-than-optimal conditions can ruin a batch of steel. "So what did you do with the hot tops after the ingots cooled down?"

"The hot tops would be removed, and any dust or debris on the surface would be blown or swept away. Then the furnace had to be turned around."

"What did that entail?"

"It had to be checked for damage to the bricks and electrodes and cleared of solidified slag; and then it could be recharged."

The thermal resistance of the refractory brick linings of the electric furnace was compromised by the fire and hot gases of repeated heats. A refractory can withstand high temperatures above 1000 degrees Fahrenheit, but at the temperatures of molten steel, 2500 to 3000 degrees, deterioration sets in, so the bricks needed periodic replacement. On a temporary basis, any pockets that formed could be filled in.

"We had a product called 'gunite,'" Mike told me. "It came in 30 to 40-pound bags which we dumped into a machine and mixed it with water. Then we had this large-diameter hose with a handle that you could 'gun in' the muddy stuff against the bricks in the furnace. But eventually the furnace would have to be taken out of service and relined."

"How often did they reline the furnaces?"

"It varied, but about once a week. After the furnaces were cooled, they could go in with a jackhammer."

"That must have created a tremendous amount of dust," I said.

"Oh yes, it was a very dirty place."

Dirty was the word that just about everyone I talked with had in their vocabulary. Simonds wasn't as dirty as the integrated steel mills that converted iron ore to steel in blast furnaces using the transformed coal from coke ovens. The electric arc furnaces and rolling mills were cleaner; however, out of the furnaces, mills and finishing operations came not only steel, but dirt and dust within which lurked troublesome particles of occupational dust diseases (asbestosis, silicosis, and anthracosis). More than one Simonds worker has suffered from these diseases. The workers breathed in microscopic asbestos fibers while they were doing hard-core work in a dust-filled environment. Heaviest exposure had come from making and removing hot tops and repairing and replacing the asbestos furnace linings and insulation.

Iconic actor Steve McQueen was in the Marine Corps before he began his Hollywood career. As punishment for going AWOL in 1949, he was assigned to clean out the hull of a ship at the Navy Yard in Washington, DC. He and some sailors who were in a similar predicament busted out all of the pipes which were covered with asbestos insulation. McQueen remembered the dust-filled air. No one wore masks. "All we could do was breathe it in," he said thirty years later when he was dying of cancer.

Asbestos was the primary ingredient of the back-up lining at Simonds at that time, but almost every worker in the plant involved in the steelmaking process was exposed to asbestos to some degree: the melt crew, the rolling mills, the shops and ancillary departments. Practically everyone who walked through the place wore asbestos-containing protective clothing: gloves, aprons, coats, and coveralls. McQueen's passion for racing cars, for example, led him to take countless breaths through the high collars of his asbestos racing suit. Inhaled asbestos fibers pass through the lung's airways and deposit in lung tissue, causing the scarring disease asbestosis.

Today, asbestos is considered a carcinogen. It is linked both to lung cancer and a tumor of the lung's pleural surface called mesothelioma, an incurable malignancy that killed McQueen and thousands of others. But unlike the radioactive contamination at Simonds, the asbestos problem was not a failure of the government or the company to protect the health of the worker—in the sense of "knowing better"—until the mid-twentieth century. Although asbestos-related diseases had been known to pathologists from the days when Simonds first began operation, its scarring and carcinogenic properties were not well known to the medical profession in general, or to the leaders of industries, where it was used until around 1950, when X-ray evidence of lung injury was linked to asbestos exposure. The word spread slowly.

Asbestos-induced pathological lung injury was an area of active investigation in the 1960s during my residency in the Department of Pathology at the University of Michigan. We scraped and digested a lot of autopsy lung tissue looking for asbestos fibers. One of my professors, Dr. Bernard Baylor, and a fellow resident, Dr. Tom Dicke, subsequently published a paper, *Prevalence of Asbestos Bodies in the Lung at Necropsy*, which, along with many other contributors in the 1960s, helped to establish an etiological link between asbestos and lung cancer.

In the 1980s a major lawsuit was filed by Peter Angelos, the first attorney to sue in court on behalf of the steelworkers. Many of them benefited, and asbestos litigation made Angelos a wealthy man. Among his acquisitions were the Baltimore Orioles baseball team. Other lawsuits followed, but cash settlements dried up with the bankruptcy filings of steel mills and asbestos companies.

A 1989 attempt to ban many asbestos products was overturned by a federal court; those products remain in use today. Workers who consider their injuries asbestos-related can now seek compensation through an "exclusive remedy" provision in workman's compensation laws. Personal injury law offices around the country regularly use the print and television media to solicit workers with possible asbestos-related diseases. Not all mesotheliomas are related to known asbestos exposure—there can be unknown exposures, and information is emerging about non-asbestos causes.

■　　■　　■

Mike said that in the old melt shop, when work was booming during the war and post-war years of the '50s and '60s, three separate crews manned the three electric furnaces on all three shifts. At least two of the three were always in operation while the third was being re-bricked, by then with non-asbestos-containing silica bricks. Because of the expense of electricity to make steel in the EAF, only one furnace was operated at Simonds in its later years, and then only on the 11–7 shift to take advantage of off-peak electric rates. In the 1970s, when business had dropped off and the electric bill was more than a million dollars a month, the costs of the morning power surge were avoided by sticking to a third-shift-only melting strategy. Simonds was a major consumer of electricity in Western New York, second only to General Motor plants in Tonawanda and Lockport.

Simonds consumed considerable electrical energy while outputting a great deal of human energy making steel. Containing the leftover energy during rest periods and after work sometimes became a challenge for fellow workers and management.

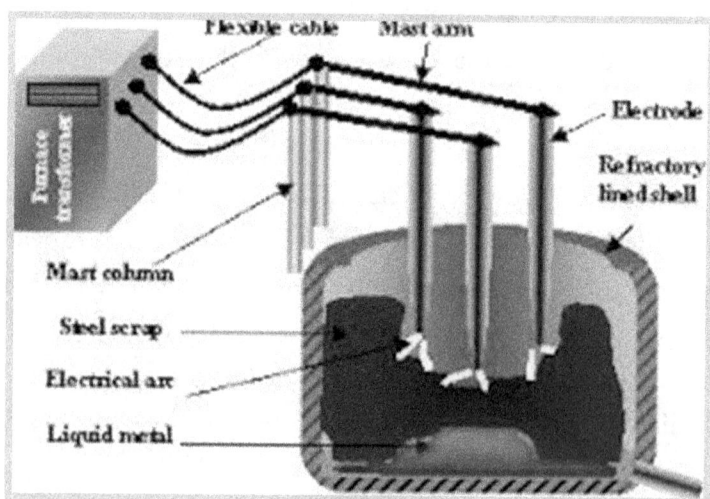

Schematic of an electric furnace. Not shown is the asbestos fire-resistant investment of the ceramic refractory layer and the outer metal shell.

9
EXTRACURRICULAR ACTIVITY

"**W**hat else do you remember about your years of work at Simonds, Mike?" It was a question I put to every steelworker I interviewed.

"It was a good place to work," Mike O'Donnell responded. "The guys helped each other out. It was like a big family. We were on tonnage, and most foremen were easy to work for. If you got your job done, no one would bother you. Of course, there were different personalities, and arguments and fights would occur. Sometimes a crew would become upset if they had to work with a fill-in who bid for but didn't know the job, but most of the time the guys covered for each other."

"In what way?" I asked.

"Well, as an example, if someone was ill or not performing well, fellow workers would pitch in. Sometimes a worker would come in drunk. But the crew would have him sleep it off while they did his work as well as their own."

"Was that common? Was there a culture of drinking at Simonds"?

"It was not uncommon for workers to camouflage their drinks and sneak them past the guards at the gate. Your grandfather was probably just one of the Italians who had wine in their lunch pail thermos bottles. Every department had a refrigerator. The workers brought in their lunches and stored them in the fridge, sometimes steaks and chops that they cooked on the reheating furnaces by their mills," he said. "And you could find beer hidden in the fridge with the food too, which was against the rules. Management tolerated it as long as it didn't interfere with the work. Beer could be found anywhere in the plant. One guy hid his stash of Genesee Cream Ale in the bar mill water baths used to cool tongs and brooms. The problem with that was that it was often pilfered. Lots of six-packs were thrown over the fence in the summer especially. I remember one incident on the evening shift on New Year's Eve. Most of one particular crew brought in drinks to celebrate, but one of the workers didn't drink. But he kept being goaded by his fellow workers to have a drink. Finally he relented, but he didn't stop with just one shot, and drank a half bottle of whiskey. Then he went berserk, shouting and gesturing wildly, and was uncontrollable. He climbed up to one of the overhead cranes and kept shouting. No one, including the plant security, could get him down, and the Lockport Police were called. It was a terrible incident that made the news and led to greater attempts to enforce the rules. There were warnings and firings, but it never stopped, of course."

Drinking and gambling were part of the steelworker's persona in every American steel mill. "I recall other stories told to me about Friday nights in one of the departments," I said. "The crew would bring in all kinds of booze: hard cider, beer, wine, and whiskey. Some of it was smuggled past the guards in bowling ball bags."

"Yes, and if they didn't drink it all at work, they would go to one of the guy's home after work and finish it," Mike said.

Work on the rolling mills was characterized by intense activity followed by desultory periods when steel was being reheated. Mill hands would play cards, go to the cafeteria, or wander aimlessly between heats or roller changes. It was during one of those extended periods on a hot and humid summer night when one of the guys called a friend at home and asked him come to the plant and toss a six-pack of beer over the fence.

Louie Koel related the story. "The friend said he just had a case of bottles. 'Okay, bring that,' he told his friend. There was little ambient light in the stockyard; only a faint glow reached it from street lamps on Ohio Street when the two men met up in shadows on either side of the fence. The case of beer was thrown over the fence, but it landed directly on the guy's head, knocking him out cold. The security guard, who heard the crash of broken glass as the bottles toppled out helter-skelter from the cardboard case, came running to investigate. "What's going on?" he asked the worker who was sitting on the ground regaining consciousness. "I don't know," he responded. "I just came out here to cool off and take a leak, and someone threw a case of beer over the fence, right on my head.'"

I thought the story was comical, and Louie couldn't stop laughing during the retelling. Some stories remain funny even if you've heard or retold them often, and they seem undiminished with tincture of time. Louie told me several Simonds tales, but I suspect this was a favorite.

"I never heard that one, but I'm not surprised," Mike said, chuckling. "There were a lot of young single guys who were feisty and drank quite a bit."

"And then there was Little George's," I said. Most steel towns had a number of bars where men would meet after work for a drink. Actually, all factory towns had one. Little George's was a favorite of Simonds workers. It was one of the closest bars to the plant, the last opportunity for a drink before work, and the first opportunity after work. But there were others in the west end, such as Freddie Taco's place across the street from Little George's, the nearby Red Robin and Summit Street Grills, and before any of those, the Rex Grill on West Avenue. The bars changed owners periodically, so different generations of steelworkers encountered different drinking environments. Sansone's in the 1950s followed Norm's, which had succeeded Grossi's Restaurant—first established on the corner of Bright Street and West Avenue in 1948 by Dominick Pasquale "Pat" Grossi while he was still a steelworker. Jim Sansone told me his father Joe borrowed $10 thousand in cash every Friday from a friend on nothing more than "a handshake," in order to cash the checks of Simonds and Harrison workers who showed up for a drink after work. On Mondays he regularly repaid the loan.

When the whistle sounded at the end of a shift in the 1940s and '50s, many men would stop first at George Taco's place just up the rise in the road at the corner of Ohio and Steven Streets. It was the working man's "Cheers," a place where "everyone knew your name." I know my father often stopped there to have a beer and to play cards. I remember some seriously heated card games at Little George's when my mother sent me over on my bicycle to fetch my dad for supper. Some steelworker wives saw Little George's as disruptive to their family's life. It may have been for some, but in my recollection it seemed to be only an occasional annoyance for my mother or me.

"It was a place where men would blow off steam," Mike said, "and when that occurred during a card game it could get intense. There are a lot of Little George's stories. George himself liked to play cards, and if he was at a table when you went into the bar, you would have to wait until he was finished with the hand before he would go back to the bar and serve you."

"I remember some pretty intimidating card games," I said, describing a man with narrowed eyes and a furrowed brow staring across the table at his remaining opponent, who sat uneasily like Clint Eastwood with a Toscanini stogie-clenched jaw. "How's this?" he snarled, as his fist slammed a card to the table, challenging the previous discard and rattling the half-empty beer glasses. Standing next to my father during some of those angry exchanges caused me to back away to the shuffleboard or wait outside.

"There were games like that," Mike said. "Not always friendly, but most were."

Tom Fiegl only worked eight months at Simonds in 1961. His grandfather, several uncles and cousins had been Simonds steelworkers. I asked him what he liked most about working there besides the fact that it was a good-paying job.

"The camaraderie was fantastic," he told me. "If I hadn't been laid off and hired on elsewhere, I would have never left." Tom, who went on to a long career in law enforcement, recalled that at the end of his first week, Bill Oates, his boss on the sheet mill, invited him to join the crew at Little George's for a drink after work. When they all assembled, Bill told Tom, "You're the new guy so you have to buy the drinks for the gang. We drink boilermakers." Tom, eighteen years old at the time, didn't know what a boilermaker was. As George Taco lined up a half dozen or more draft beers and poured a shot of whiskey for each glass, Tom sat slack-jawed on the barstool silently contemplating what it was going to cost him. "I really loved working there," he said.

"It was the best place, a beautiful place to work," Al Ferrante said. "Everyone was a friend and the bosses were good men. Don Richards was a good superintendent. I liked the work. I worked at Harrison's before I went in the Navy, but I didn't want to go back. I liked the routine at Simonds, twenty minutes of hard work and then twenty minutes of rest. You could go get a cup of coffee or play cards. It was fun. And later, when I worked in the lab, we had time to order in pizza or doughnuts, talk and enjoy ourselves between tests."

Dave Craine echoed those exact sentiments. "It was a fun place to work. I know that a lot of guys waking up in the morning dreaded going into work, but I couldn't wait to get there," he said. "There were a bunch of clowns and we played practical jokes between the heats when we weren't playing cards."

Most every steelworker I spoke to about Simonds felt good about themselves and were as proud of their work as Mike, Tom, Al, and Dave were; but not everyone felt that way.

Anthony "Doc" Ruggeri was an exception. I had spent two eight-hour evenings with him in the 1950s driving to and from New York City with the Pipitone family to visit relatives. He was an incredibly humorous storyteller on that trip, mimicking voices and impersonations of his subjects that kept us in stitches the entire time. I called him sixty years later at his home in Boca Raton, Florida where he had retired, and asked him about his work experience at Simonds in the early '60s..

"It was a terrible place," he said. "I don't know how I worked there for three years. I don't know what hell is like, but I hope I never go there, because it must be like Simonds." I considered Doc's point of view—if one could weave all of the Simonds sights and sounds into a single tapestry: hard labor by fiery furnaces, smoke and steam, hammering noise and acrid smells—it could represent someone's vision of hell. "I don't know how guys worked in some of those places," Doc continued. "I went into the Pickle House once and I couldn't breathe. There was a guy who worked in there that wheezed all the time. It was awful. I had a cousin who worked in the cast magnet department. He asked me to come see him when I took a break. I opened the door to the room where he was working. It was pitch black. I called his name a couple of times. I guess he finally heard me because he came out of the darkness like someone coming out of a tomb. The only thing I could see was the white of his eyes."

John Coleman found that story believable. "He could have been talking about jack hammering out the lining of the Ajax furnace or sandblasting," John said. "We did have a small enclosed space where sandblasting was done to remove scale. It could be like a sandstorm in there at times and the worker had to breathe in that dust, but it's more likely he was talking about the furnaces." The foundry furnaces, like the larger electric furnaces, had a refractory lining, but it was a brown asbestos sheeting, not bricks. The lining required periodic removal and replacement. They also employed hot top seals which added to the dense fog of dust when they had to be removed and blown with the air hose. John described what his clothes looked like at the end of a rare shift on that particular job.

"In the locker room I stripped down and dropped the clothes on the floor and took a good shower. Then I picked them up where I left them, wrapped them in my towel and put them in a shopping bag. There was no way I could get into my car or my house with them on. I hosed them down in the back yard, then soaked them in clean water for a few days. Then I could put them in the washing machine for a separate cycle. The dust on my shoes wore off in time."

For Doc, the heat in the plant was the worst part, "especially in the summer," he said. Lockport was suffering through a dry, hot summer when he worked there. Weeks without rain, parched fields, and brown lawns were an unwelcomed weather pattern that periodically visited Western New York. Doc continued, "We had an Indian from the reservation who worked with us on the sheet mill. During one of the breaks on the night shift I said, "Joe, why don't you go out and do a rain dance?" No one considered the insensitivity of such a remark back then. "It was clear that night, the stars were out," Doc went on, "but he went out and came back a few minutes later smiling, 'I did it.' We went back to work and about two hours later we heard a tremendous downpour hit the siding and roof."

Doc's impression of Simonds was far from typical, but Doc wasn't a typical steelworker. Nearly everyone would pick him out of a lineup of steelworkers as the man who didn't belong. He drove to work in a Cadillac, dressed in fashionable slacks and sports shirts, his long dark hair styled and carefully combed to set off his black, horned-rimmed glasses. In the locker room, he changed into his shapeless gray work clothes. At the end of his shift on the sheet mill, he followed the crew in their sweat-stained, sour-smelling clothes back into the locker room. After showering, he changed back into his snappy street clothes and clocked out, often accompanied by a friend who had similar tastes in clothing and drove a Lincoln Continental. "Where you fellas think you're going all dressed up, to a party?" fellow workers would chide as they made their way to their cars. When Doc got home he would invariably wash the Simonds soot off his Caddy.

By the time he reached adolescence Doc had developed a passion for clothing: suits, shirts, slacks, and ties. Dress at the high school and college level, the "Joe College" look, was a popular style in the '50s. Lerch and Daly Men's Shop had been established years before Lockport was even a twinkle in Dan Simonds' eye. Albert's Men's Shop was next to the Rexall Drug store where I worked in high school. Albert's a stylish line of clothes that Doc might have worn. But he didn't want to dress like anybody else in town, so he stopped shopping in Lockport's men's shops, preferring the clothiers in Buffalo.

Workers at Simonds were paid well in the late 1950s; at one time, they even enjoyed better salaries and benefits than employees at Harrison's. Doc made good money and he spent it well, appreciating as others did the financial rewards of double shifts. One winter night after completing a double shift at 11:00 p.m. on the sheet mill, a raging snowstorm prevented workers from driving home, and the 11–7 shift from showing up to work. So Doc did a third shift. He would have continued with his normal 7–3 shift the next morning, but the snow had stopped and snow plows had cleared the road. His boss had arrived to find him ready to start work. "What the hell are you doing here?" Johnny Apolito yelled. "Go home!" When the paymaster's armored vehicle pulled into the plant that following Friday, Doc found his pay envelope was thicker than usual. He had recently married, and that night he went out and bought an entire set of appliances. Into the hand of a stunned west end appliance salesman, Doc rolled out bills one after another from his wad of cash.

Simonds' wages provided a middle-class lifestyle, but it was definitely not where Doc wanted to be, even when performing the lighter mill jobs. He went on to get a junior college degree from Erie County Tech and then a job in the Buffalo-area steel industry. Decades later, retired in Florida, he developed cancer, a form of non-Hodgkin's lymphoma, one of twenty-two types of cancer the federal agencies deemed eligible for a settlement. One of the first things he said over the phone before I asked the first interview question was, "Do you know I got a settlement?" It was a check similar to the one my mother received for my father's cancer-related death.

Like Doc, other men left Simonds to work elsewhere, such as Harrison's. Some were sorry they did and went back to Simonds. The Harrison environment was different from Simonds, much cleaner and quieter, more organized and regimented. Some said it was less friendly, but Harrison's had more than their share of loyal workers. There were skilled workers at both plants, but many of the assembly line jobs at Harrison's were monotonous. I know: I worked there one summer pulling air conditioner units from an assembly line and combing the bent cooling plates straight again. Simonds was physically demanding, but I was never bored except for the times I did Harrison-type work in the sheet mill inspection area.

There were social and cultural differences as well, exemplified by the company picnics. The Harrison annual picnic was a family affair not to be missed. It was always a big event in Lockport every summer. A few thousand men and women employees, spouses, and children filled the grounds of Krull Park in the village of Olcott Beach on Lake Ontario. It was a day-long event with an allotment of tickets for a variety of food, drink, and carnival games with prizes and souvenirs. Children whose parents were not Harrison employees, like me, would tag along with friends whose parents were, happily sharing their tickets. By sundown it was over.

The Simonds picnic was a smaller, relatively quiet affair initially held in a small glen at the west end of town, the Hickory Club, before it moved to the Niagara County Fairgrounds. It was co-sponsored by the union and management. District union officials were invited, but usually demurred. The event was attended only by active and retired steelworkers, although plant managers would make an appearance. John Rinaldo and Elmer Gagliardi, former steelworkers who had opened a bar and restaurant in town, catered the affair. The menu included "beef on weck," burgers, dogs, and beer. They also shucked a lot of clams, grilled a lot of corn, and sliced a lot of watermelon—popular items around Western New York in the summer. The games were craps and poker, or for the retired Italians, such as my grandfather and his friends, the energetic finger-counting game, *morra*. There were no women at the Simonds picnics even though women had salaried positions in the office, and children's attendance was discouraged. I went to one or two of the Hickory Club events with my grandfather when I was a teenager. He enjoyed the reunion, drinking and chatting with his old friends and coworkers. By the end of the day he was feeling pretty good, and my grandmother was thankful I could guide him home.

When Simonds began operations in Lockport, women had limited opportunities. In fact, it would be another decade before they got the vote, and even later when legislation was passed giving women employee opportunities and pay equivalent to men. Although women at Simonds were never a part of the hourly workforce in the mill, over the years they filled positions in administration, health and benefits, and amenities.

10
WOMEN OF STEEL

One of the reasons some of the Simonds men gave me for choosing Simonds over Harrison's was that they preferred working with men more than women. The distaff side of Simonds was pretty much limited to the cafeteria and office building. New York State labor law, however, required Simonds not to discriminate against women in their hiring practices. "If they did hire any women in the plant, they didn't last long," Mike O'Donnell said. Tony Parete recalled several women being given tours of the plant as part of the hiring process, but none of them ever returned to work there. "In twenty-nine years, I never saw one woman working in the mill," he said. "They looked around, but when they walked out, the last that anyone saw of them was the door hitting them in the ass. They went to work at Harrison's where it was much easier."

Jim Calos, the melt shop superintendent at the time, corroborated those recollections, although he remembered one woman who worked there as an assistant to the bricklayers relining the furnaces. "At the end of the first month, the woman showed up at my office. I knew immediately from the look on her face why she was there," Jim said. 'I'm going back to teaching.'"

Actually, the Buffalo area steel mills at the time, Bethlehem and Republic, a number of women workers into their mills and finishing departments under a program in which they were given the opportunity to do some of the hot, heavy, dirty, and dangerous jobs that men performed.

However, the resignation that took place in Calos' office pretty much left women at Simonds consigned to their historic roles in the office and restaurant. The origins of the restaurant at the Simonds plant in Lockport are obscure, but it piqued my interest about women's work at Simonds.

The Simonds family provided restaurants for its employees in the early days at its Fitchburg, Chicago, and Montreal plants as part of their amenities, so one would assume that one was provided for the Lockport workers. What is known is that a restaurant, often referred to as the commissary in the old days, was owned and operated for years, beginning in the 1930s, by Frank and Amelia Kinsler, who were West Avenue residents. Frank, who became blind, passed away in 1942, but his wife continued to operate the restaurant with the assistance of her daughters, Mildred and Olga, and two other women staff. Her granddaughters, Arlene Sweet and Shirley Hayden, who comprised the rather unique third generation of women workers at Simonds, pitched in after school. Breakfast and lunch were served from 9:00 a.m. to 2:00 p.m. and supper from 5:00 to 7:00 p.m. Arlene recalled the industrial environment of the plant, driving with her grandmother to help serve the dinner crowd. They were allowed to drive through the main gate at the south end of the property; and, unlike the workers who had to park on Ohio Street and walk to their departments, they were able to swing around the buildings from west to north, and park by the back door of the restaurant at the end of the building that housed the grinders, close by the bar mills. In the restaurant they would meet the staff and sometimes a few steelworkers: Joe Dumphey, Ed Mulvey, and Eddie Ryan, who volunteered their time to serve food. The men also helped stocking shelves and refrigerators, moving crates of Castle's milk, Johnny Ryan soda pop and Vernors ginger ale to and from storage. Shirley Hayden drew me a sketch of the layout as she remembered it. It had generous work areas with large gas ranges and refrigerators, a center island which doubled as a work surface and table for employee lunch breaks, sales counters, a dining area, and an upstairs storage area.

Many workers brought lunches or dinners with them from home in black lunch pails or brown paper bags. Those who didn't could choose from a variety of hot items. Arlene Sweet told me that some of the daily blue-plate specials included roast beef, pork roast, and meat loaf accompanied by potatoes and vegetables. There was lighter fare, too, like soup and sandwiches. Hot coffee poured from one of the two large urns near the cash register was the winter beverage of choice, whereas a variety of sodas were popular in the summer. The men burned through a lot of calories, and plenty of replacements were available in the form of baked goods, doughnuts, cookies, and cupcakes delivered daily from Layer Brothers bakery, and there was Castles and Sealtest ice cream.

Shirley made me a sketch from her memory of the dining area which had a half dozen picnic-style benches and tables. It was walled off from the mill, but it had an open door facing into the plant and a bank of windows that were often ajar, making the area difficult to keep clean. Arlene remembered wiping down the tables that were too heavy to move without the assistance of steelworkers. The restaurant was located "downwind" from the grinding department and Pickle House. The sour odor that drifted in from the Pickle House. was irksome to Arlene and Shirley, but the steelworkers, long accustomed to Simonds' fine-particulate atmosphere, seemed not to have noticed. Smell is a major component of taste; even the sweetness of ice cream can be blunted by a sour scent. John Coleman, referring to Simonds' black granular dust from the grinders, said, "If the pepper shaker on the table was empty, there was plenty of it lying all around you." It's also quite possible that uranium and thorium dust drifted in with the grinding dust in the early '50s.

The original restaurant was gone when I worked there in 1959. The bar mill area had been extended and the finishing department expanded, obliterating the site of the old restaurant. I attended the University of Buffalo, where food service in the cafeterias of the residence halls and student union was provided by the Cease Company, a food purveyor for many institutions and industries in Western New York. Sometime in the mid-fifties after Mrs. Kinsler retired, the Cease Company took over the food service at Simonds. Jack Frost (who could forget that name?) wasn't nipping at anyone's nose, but he was managing the Cease operation. It was a cafeteria, not a restaurant, and the site had been relocated to an area near the band and cogging mills. It was a narrow space walled off from the plant with a bank of windows. Basically it was a food line with no seating available. Workers entered at one end, slid their trays along the metal counter to pick up hot and cold items, drinks, etc., and then exited the opposite end after leaving the cashier.

Workers chose either to take their purchase back to their departments or find a place to consume it somewhere in the plant. I sometimes sat on a pile of steel ingots outside the enclosure to eat, not the most comfortable spot, but I often chose the outdoors on the more pleasant summer days. On one of those days after lunch I was sitting on a wooden box, leaning back against a wall, my eyes closed, absorbing the sun's rays, when one of my father's friends came up to me and said, "I bet you're happy you'll be going back to college in September." I nodded, but what I was thinking was that I really enjoyed the work I was doing and wasn't all that anxious to go back to school.

After Arlene graduated from high school, she worked for two years in the Lockport motor vehicle bureau and then went to Simonds, where she worked as the bookkeeper in the credit union housed in the mill office building. After the restaurant closed, her mother, Mildred, transferred to the same building where she worked on production records. Because the credit union, the first aid department under the direction of a registered nurse, payroll, personnel, and a shoe shop were located there, this building afforded interactions between the women who worked there and the workers in the plant. Her mother was one of the few women seen in the working areas of the plant. Her job required her to walk from the mill offices through the plant each day to collect production data the foreman recorded on the job. These entries were made on small wooden desk stands adjacent to a rolling mill or in larger, windowed spaces in the sheet mill inspection, bar mill finishing, cold roll, and melt shop departments. One day, when one of the workers criticized the European style of my father's numerical entries, my father shot him a cold stare, silencing him with the reminder that "the Romans invented numbers." When he told me the story I said, "Yeah, Roman numerals, not Arabic." Actually the origin of Arabic script is Hindu. Invading Arabians adapted it, and so it became associated with them, but I thought I had shared enough with my father and didn't press it. In our walk-through, Michael and I came across several small, debris-laden office and clerical spaces where the record keeping took place.

Arlene often spoke to workers who came to the credit union to borrow money from their accounts. Retired workers remembered her as a sweet and helpful woman. Some of the Italian men couldn't speak English well, she remembered, but "they knew exactly how much money they had in their accounts." The credit union was located directly across from first aid, which was staffed on all shifts, so Arlene often witnessed men rushing in or being carried in with injuries, burns, and cuts. The minor injuries—most cuts, eye irritants, minor burns, and abrasions—handled by the nurse, Donna Randolph; but the most serious accidents in the plant bypassed the first aid clinic entirely and were evacuated, sometimes by ambulance, straight to the hospital.

Reutter's book includes a chapter on women's work at Bethlehem's Sparrow Point plant where the men did the heavy physical labor, but women, called "tin floppers," worked at tables stacked with newly milled thin steel sheets. The women had the knack and hand and eye coordination to rapidly sort and flop the sheets while inspecting them for defects. Not all of the sheets were thin and light, however; some were heavy and had rough edges, and gloves didn't always protect the hands from cuts.

A chapter devoted to women in Buffalo's big steel plants of Bethlehem, Lackawanna and Republic in Regovin and Frisch's book, *Portraits in Steel,* has striking black-and-white photographs of women in hard hats and heavy industrial clothing looking intently into documentary photographer Milton Regovin's camera. Michael Frisch narrated in wraparound text the responses to his questions of what it was like to perform hard, heavy work in a man's world of steel. A young Jennifer Beal's portrayal of a welder in a Pittsburgh steel mill by day and a dancer by night in *Flashdance,* fictionalized the experience of those women.

World War II proved that women were capable of working in heavy industry when they entered the work force in great numbers, filling in for the men in factories around the country who left for military service. The steel industry had their share of Rosie the Riveters (Rose Bonavita of Peekskill, New York was the cultural icon of the movement). The women proved that they could do fairly heavy manual work during the war years and later at the Bethlehem plant in Lackawanna, where they were employed as burners, cutting bars of steel with heavy torches, rigging and banding up steel from the rolling mills, loading and unloading box cars.

During the war women had been employed in a variety of blue- as well as white-collar jobs in all Bethlehem steel mills' finishing and inspection departments, not just as bricklayer helpers, but as crane operators, welders, machine operators, maintenance workers, and pistol-packing security guards. These were jobs that women could have done at Simonds in Lockport, but no women worked in the plant during or after the war except in the office and restaurant.

SIMONDS'
The Vintage Years

Simonds Brothers, the Scythemakers, who formed the Simonds Manufacturing Company in Fitchburg, Massachusetts in 1868

Dan Simonds, president of Simonds, established Chicago plant in 1900; moved to Lockport in 1910.

Initial Simonds buildings (1&2) with gabled superstructure, going up in 1910 on Ohio Street by the Erie Canal

Producing armor steel alloy under roof at Simonds during the last year of World War I

Simonds employees participate in Lockport NRA (National Recovery Act) parade, August, 1933

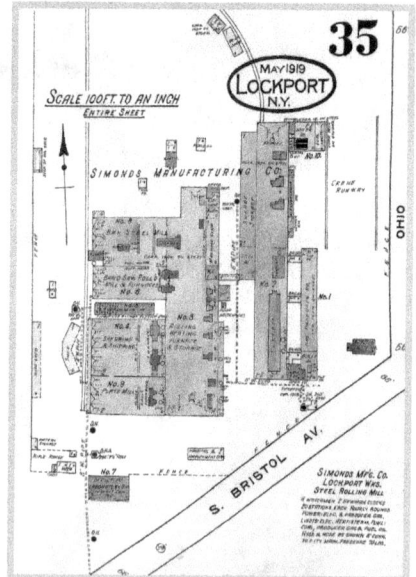

Layout of numbered buildings at Simonds after addition of rolling mill, 1919

Some of the Simonds employees, admisnistrative and workers, pose by office building, 1933

A winter scene of Simonds in the 1920s with drained Erie Canal in foreground

Simonds, 1947. Material yard with Coal and scrap

Shoveling alloy into original 6-ton EAF

Tap and pour from the 6-ton EAF into ladle

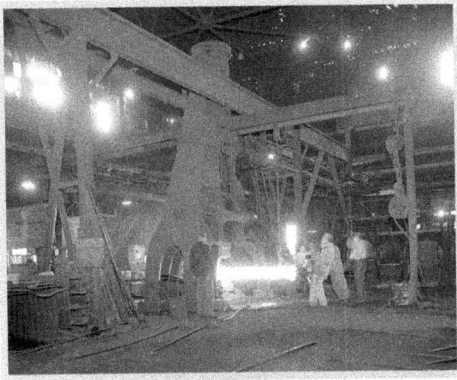

Steam hammer crew manipulate a hot ingot.

Working on a bar in the bar mill finishing dept.

Worker on left turns the screw while rollers finish a sheet.

Sheet mill crew taking a break between heats. My father leans in to center.

Sheet Mill Rolling and Finishing Department, 1958. Anthony Rosati, front row, 3rd from left.

16-inch bar mill crew hooking up, leveraging and rolling a bar as broom sweeps remove scale.

Cogging Mill worker grips and rolls a hot steel plate.

Band Mill. Two workers lift a long band of steel into a set of rollers driven by the a rope/turbine arrangement.

Band (Rope) Mill. Workers walking a hot band back across the partially dry wood plank floor.

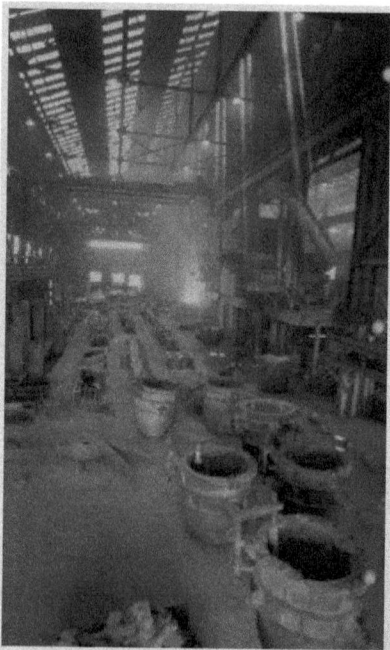

New melt shop with furnace ladels. Tap and pour from 15-ton EAF in the background.

Cover photo. The tap and pour from the 15-ton EAF. First Helper and assistant watch from platform.

SIMONDS'
Abandonment and Decay

The workers' entrance and security office. The fenced-in excised property is becoming enveloped by vegetation as broken windows, fallen roof tiles and corrugated-metal panels rust and collapse, exposing the interior to rain, the fallen leaves of autumn and the snows of winter.

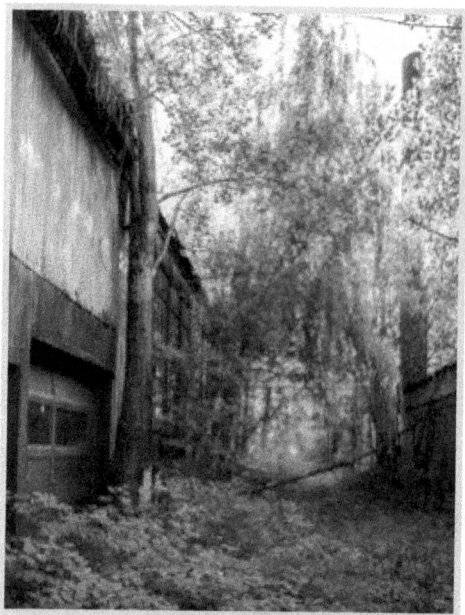

Nature is reclaiming the site.

Fresh winter snow falls from the open roof panels.

The hollowed-out empty chamber echoed as we walked and talked.

A winter perspective of another long, empty interior space

The gantry cranes that once moved tons of material and steel product now rest.

The Fletcher's job was to repair the hemp rope.

Rope and turbine drive for the band (rope) mill

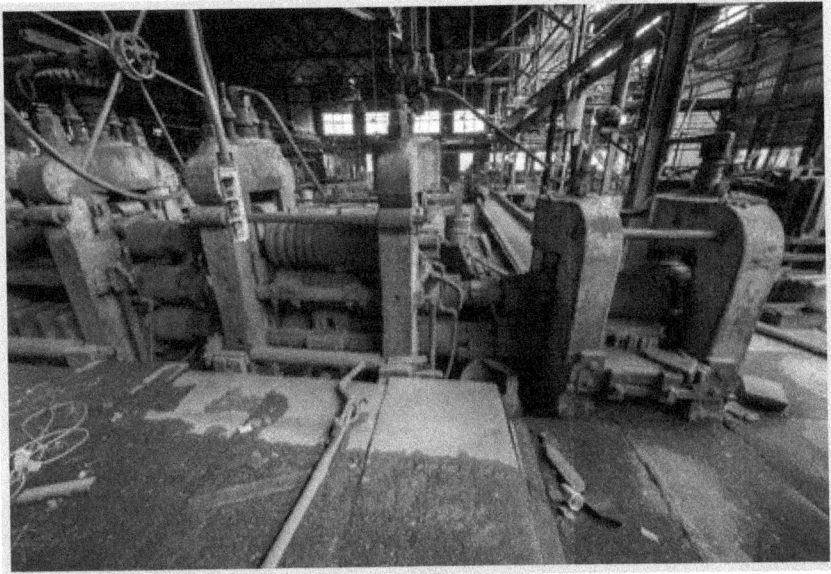

The 16-inch mill stand left behind at auction

A rusting annealing oven sits on a moss-covered deck.

Pat's shop littered with implements once forged there for use throughout the plant

Silent sentinels in winter

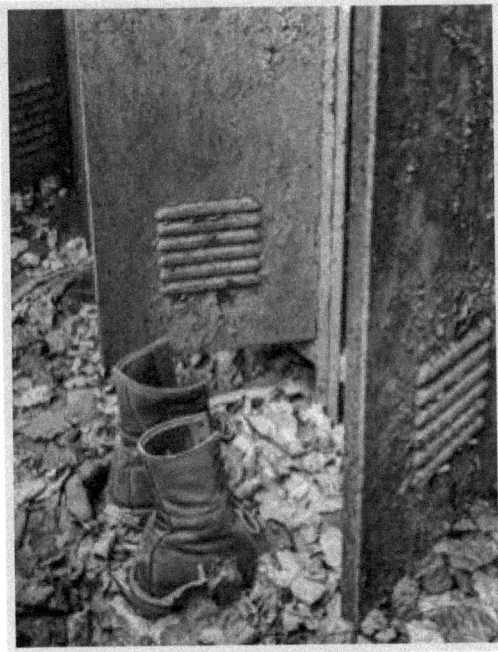

Boots, fallen leaves, and a rusting locker

Dick Sadlocha's hat, among the things left behind

A montage of the Pickle House images, where steel product was acid-treated after rolling

A steel mill space becomes a grow house.

Part of the record and document dump in foundry at Simonds.

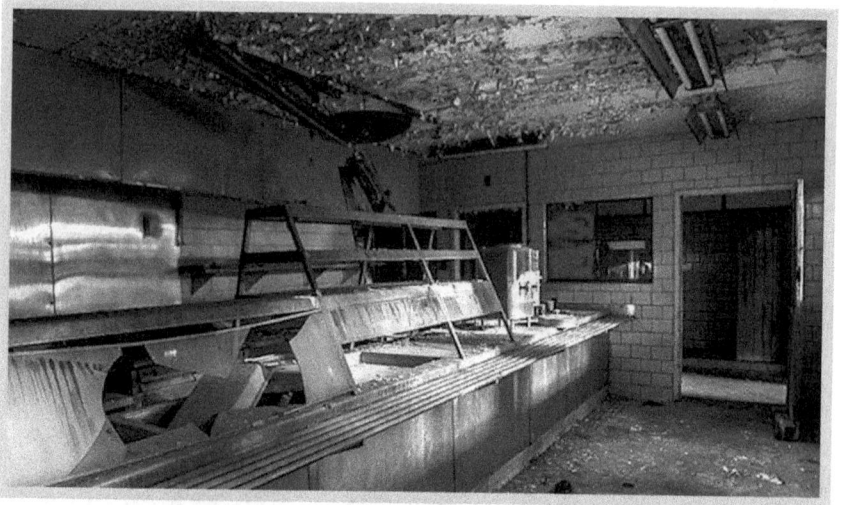

Cafeteria: Where have you gone, Mrs. Kinsler?

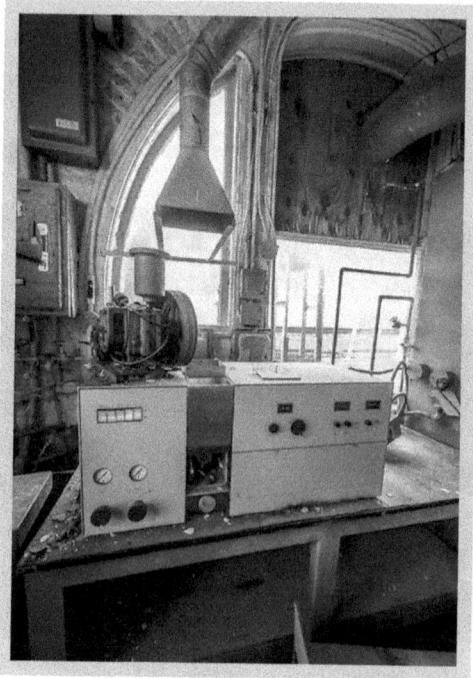

An old analyzer gathers dust on the second floor
abandoned Chemistry Lab in the Office Building.

Aerial view of Lockport's west end with Simonds built out in the 1960s. Erie Canal at right foreground edge.

The Simonds Superfund site, August 2018. Nature is reclaiming the Simonds buildings with the missing roof tiles, and the old employee parking lot in approximate center. ATI building on far west side, Erie Canal in lower right, and Lockport's west end to the north, fading to the Niagara Escarpment and Lake Ontario on the distant horizon. Photo: Sky Link Drones, Williamsville, NY.

11
SIMONDS GOES TO WAR

When Simonds began operations in Lockport in 1910, the country was at peace. Problems on our southern border soon developed because of the Mexican revolution and the conflicting ambitions of the rebellious Pancho Villa and the Mexican government. However, our military involvement was pretty much limited to small-scale skirmishes, until General John Pershing led a "punitive" expedition into northern Mexico in response to Villa's attack on an American border town.

By 1915, however, most of Europe was engaged in World War I; but the United States, under President Woodrow Wilson, attempted to maintain neutrality despite pleas from Britain, one of America's closest trading partners. Nevertheless, American business and industrial interests found ways around congressional neutrality acts and surreptitiously shipped supplies to the British and French.

In 1915, the Germans began unrestricted warfare, sinking merchant vessels from several countries, including the United States. The passenger ship *Lusitania* was among the vessels sunk. Among the 2,000 passengers, over 1,000 were killed, including 128 Americans. It was later learned that the *Lusitania* was more than a passenger ship. Its hold contained thousands of rounds of Remington ammunition bound for England. Two years of protests, apologies, and more sinkings followed. When President Wilson's final warnings about the renewal of unrestricted submarine attacks on our merchant ships went unheeded, the United States declared war on Germany in April 1917. It took a year to mobilize, recruit, train, and ship an army to France. The American Expeditionary Force under General Pershing was finally inserted at a critical point of the war, taking part with allies in the action around Chateau Thierry, Belleau Wood, and the final Meuse-Argonne offensive, which led to the armistice on November 11, 1918.

At the war's outset, the United States was the world's largest steel producer, and during the mobilization Simonds joined the area's integrated steel companies supplying steel to the military. Gearing up for war production was rather chaotic for most industries. There were so many labor and management issues over supply and demand that the Wilson administration established the War Industries Board to mediate disputes, coordinate the purchase of war supplies, and encourage industries to eliminate waste and maximize efficiency. Productivity increased, especially in those plants employing Taylor's management methods.

Simonds' 30-by-90-inch plate mill crew rolled nickel and molybdenum alloy armor plate for the U.S. Naval Bureau of Ordnance. Molybdenum and tungsten's ability to maintain steel's hardness at high heat made them important military metals. Simonds' armor plate was especially selected for artillery units, including a massive rail car-mounted gun. The company also designed steel helmets and thin armor plate for locomotives and tractors. Simonds' reputation for quality steel and rapid delivery increased demand for its products, and it responded by expanding its facilities and work force in 1918. But while the deliveries from American industries in general came rather late, the post-war industry benefited from the mobilization and productivity gains.

To meet the post-war demands, Fitchburg operations were consolidated into one large facility, the world's first modern windowless and climate-controlled plant. Conceived and designed in the 1920s, construction began in 1931; however, the Depression would force the company to delay bringing that and its other projects to completion until the end of the decade. In 1922, the Simonds Manufacturing Company incorporated as Simonds Saw and Steel. Around that time Simonds purchased an abrasive company in Philadelphia. Products to cut, grind, polish and sharpen cutting surfaces were an essential component of the specialty steel process. Subsequently renamed the Simonds Abrasive Company, the Lockport plant would become one of its major end users.

Along with plant expansions and acquisitions, Simonds added workers, many of whom were immigrants such as my maternal grandfather, Antonio DiPaolo, his cousin Luigi DiPaolo, my paternal grandfather, Luigi Rosati, and his brothers, Francesco "Frank" and Vincenzo "James, aka Big Jim" Rosati. The Simonds organization made a point of publically declaring their human resource policy of "selecting men eminently fitted for steelmaking," and followed through by hiring thousands of European immigrants. Production, not ethnic origin, was the Simonds priority.

My maternal grandfather, Antonio, had immigrated to America with his father and younger brother, Camillo, before the war. The men were part of the great European migration that had begun in the mid-nineteenth century, a movement that would slow to a trickle in the middle of the Roaring Twenties.

In the case of Italy, the regions and city-states had only become unified as a nation in the 1860s. The unification was a mixed blessing. It created a state, but the fractured society of the industrial north and agrarian south created conditions of enhanced impoverishment, especially in the south, leading to the exodus of millions of its citizens.

Like other Italians, my grandfather, his father, and and brother came to America to work and earn money to support the families they brought with them or left behind. In my grandfather's case, his mother, four sisters and a younger brother, Bartolomeo, a future Simonds worker, remained in the old country. Such American values as freedom, independence, and equality were secondary considerations. Most came from provincial towns and villages or the outlying areas of the larger cities like Naples, Reggio and Pescara. The state wasn't that important to most of them; in fact, there were deep divisions in terms of direction and visions for the country. In that context, many of the impoverished residents of the small towns and villages could be described as *campanilismo* (from *campanile,* or bell tower), their loyalties extending no further than the sound of the church bells in their village. In America, however, those attitudes changed and patriotism emerged even before full assimilation. "You're an American now, speak English," was a phrase often heard in the neighborhood where parents dressed their youngsters in Army and Navy uniforms during the war years.

Initially my grandfather, his younger brother, and father found work on Pennsylvania's railroads as manual laborers, "gandy dancers" who lived in shanties by the side of the tracks. When his father became injured on the job, my grandfather was forced to accompany him back to Italy, where he got caught up in the Italian army during the war with Turkey and subsequently World War I. My great-uncle Camillo, who stayed behind, entered the American army in 1917, and with Pershing's expeditionary force fought in France, while my grandfather took part in Italian army operations in the Veneto region of Northern Italy against the Austrians. After the war, my grandfather, who was now married and had two children, returned to the U.S. He reunited with Camillo, who had returned from the war, married and taken up residence in Buffalo. Relatives from my grandfather's

Italian village of Roccamorice and my grandmother's nearby village of Lettomanopello had preceded my grandparents to Lockport. So, following the common pattern of chain migration, they proceeded to settle in Lockport's west end.

My grandfather was hired at Simonds and began a long career, starting out on the labor gang, then the rolling mills, and finally, when I was old enough to comprehend his job, the bar mill finishing and inspection department from which he retired. His story was not unlike many other immigrant Italians from Roccamorice and other villages in Abruzzo and Molise, Calabria, Campania, and Sicily who settled in Lockport's west end and lower town, who found work at Simonds. For many, the steel mill seemed a dirty and dangerous place, so different from the simple and picturesque farming villages they had left. Some would have gladly taken the boat back to Italy but for the realization that the towns they left could barely support the residents who remained.

I recognized scores of those men in group photographs of the melt shop, rolling mills, steam hammer, cleaning, and inspection departments taken in the '40s and '50s. They were scattered among the rows of men of various nationalities and ethnic backgrounds who had immigrated or migrated to Lockport under similar circumstances. Most of them had spent years on the job; some advanced to shift foreman, a few to mill supervision, but none that I knew were promoted to senior positions in the administration. Their modest homes were tucked among the businesses on West Avenue between Hawley Street and Crosby Avenue that included several "mom and pop" bars, bakeries, restaurants, and groceries, mostly with Italian surnames. There were other commercial enterprises in that neighborhood, such as drycleaners, a glass factory, furniture, and appliance shops.

The greater neighborhood included Park Avenue, which paralleled West Avenue to the north along the New York Central Railroad freight yard.

Park Avenue was intersected by Crosby, Bright, Michigan, Ohio, New York, and Bristol Streets, the streets across which I would lead kids as a member of the Charlotte Cross School Safety Patrol. These streets encompassed a vibrant, nurturing working-class community where steelworkers once lived, and where, sadly, now they and the families they left behind have all but disappeared. Today, the bleakness of that working-class neighborhood reflects the decay of the abandoned plant on its southern edge.

Steelworkers and other union wage earners spent most of their money where they lived and worked. The community's loss of those men and their wages, left many neighborhoods as hollowed-out and shabby as the abandoned buildings that remained.

Simonds was a short walk for many men, and some lived close enough to return home for lunch when the steam whistle sounded at noon. Sicilian immigrants who worked at Simonds but lived in Lower Town, the site of Lockport's earliest and largest Italian enclave (Lockport's early history began above the Niagara escarpment, but after the dig and placement of locks, it continued at the escarpment's base) had a much longer walk, which could be tough on those Western New York winter mornings, when snowfall and wind chills hovered around zero and below. The same was true for Simonds' rural employees. One steelworker's daughter remembered her father, Merrill Ranke, walking to his job at Simonds in the 1940s from their country home along wind-drifted snow banks that reached halfway up utility poles. "Snow days" were a rarity at Simonds. When men working double shifts couldn't make it home, family members would sometimes deliver meals.

My grandfather was a walker and refused my father's offer to pick him up for work even on the meanest winter days. I learned his story from my mother, who began writing a memoir in the last years of her life. Her story of a steelworker's daughter began as a five-year-old coming to America in 1921 with her mother and seven-year-old sister, reuniting with her father, who had found work at Simonds. He settled his family into a two-bedroom apartment in Lockport's west end. The residence was attached to the Red Robin Bar on the corner of Ohio Street and West Avenue. My grandfather divided the girls' bedroom with a rope and sheet so that he could rent space on one side of the curtain to two steelworkers who worked different shifts. That arrangement, not uncommon among immigrants in the 1920s, would seem outrageous today. A few years later he had saved enough money to buy a house down the block at 353 West Avenue. In a home between those two residences on West Avenue, my grandfather would one day join other workers to form a labor union.

My mother's memory of the house where she grew up during the Depression was especially vivid as she recalled the drudgery of washing and ironing the grimy sheets, towels and work clothes of the steelworker boarders they housed. Deborah Rudacille, in her memoir, *Roots of Steel*, suggests that women such as my grandmother and mother, the wife and daughter of a steelworker, were connected to the steel mill by their domestic labor, as were many other women in the neighborhood who had boarders in their homes. In those days, boarding steelworkers was not uncommon, and for many it was a win-win. Many Lockport steelworkers found their first residences as boarders, and the rent they paid enabled families to supplement their steelworking income and pay their mortgages. Those boarding homes and working at Simonds gave many immigrants the foothold they needed to establish themselves and their families in Lockport.

Steel mills across the country came close to complete shutdowns during the Depression. Simonds had to tighten its belt in the face of reduced demand for its steel products, but the company tried its best to maintain a workforce and kept layoffs to a minimum. However, many departments only operated part-time, and in general, throughout the plant the pace of work significantly slackened. Working hours were erratic and unpredictable. Often men would show up at the plant not knowing if they would work that day, and many men feared losing their jobs or not working enough to support their families. In the days before the union and employment agreements, how one got hired, how often and the hours one worked, including the foremen, was dictated by the plant manager and bosses, who were all-powerful. A Simonds owner or executive from Fitchburg on a visit to the Lockport plant would think nothing of pulling a foreman into an alleyway, grabbing his shirt collar and threatening physical harm if he was unhappy with the steel being made or rolled. Physical altercations in the plant between workers and between supervisors and workers were not uncommon in those days. "People got away with behavior they would be fired for today," the old-timers recalled.

Beyond his production value, a worker's worth could also be determined by whatever else he could bring to the job besides his labor. A good winemaker or his wife's pasta sauce might get the worker more shifts. Among the plant's workers were local farmers who favored the boss with produce in an attempt to remain in good graces and stay employed during the long stretches between growing seasons. Tony D'Angelo was the boss of the labor gang and made daily or weekly assignments from his pool of workers in the 1930s and '40s. When the assignments were all made, anyone left was sent home. By all accounts Tony was a well-liked and respected boss, but he was something of a legend to later generations of steelworkers who recalled hearing stories of workers bringing him baskets of fruit and vegetables, hoping to be assigned paying jobs for the week.

■　　■　　■

The immigration and Depression decades of the 1920s and '30s coincidentally witnessed a period of anti-war sentiment that permeated the country, including its steel mills. A reflection of the pacifism was the signing of the Kellogg-Briand Pact renouncing war as a means of gaining international power. That was followed by the passage of three Neutrality Acts (the last over President Roosevelt's objections) prohibiting arms and materiel shipment. However, surreptitious shipments to Britain took place, just as in WWI, and the last Neutrality Act established an overt policy of "cash and carry."

When WWII broke out, the country was better prepared than it was in World War I, thanks to a war mobilization plan that the U.S. Army had developed. Following the Pearl Harbor attack, the plan allowed President Roosevelt to, almost overnight, transform American industrial might into the mass production of war materiel, as the country became an "arsenal of democracy." Even before Pearl Harbor, the government was signing defense contracts with steelmakers, whose purpose at first was to fill requests for steel orders from Britain. In May 1940, Winston Churchill cabled President Roosevelt requesting ships, planes and steel. Roosevelt, before the 1941 Lend-Lease Act, said "no" to the military hardware, but "yes" to steel. In November, he had a photo-op tour of Western New York defense industries, including a visit to Bethlehem Steel in Lackawanna, which had become a formidable producer of high-carbon steel. Bethlehem would roll out miles and miles of it during the five war years. Lockport did not make the president's tour, but Simonds geared up to supply its special alloys to the war effort at the same time its workers were organizing the local chapter of the newly formed United Steelworkers Union.

Steel armor-plate alloy from Simonds found its way into various armored vehicles, including landing boats used in the Pacific campaign against the Japanese. There was a limited number of these craft at the beginning of the war. One of the limiting factors was the availability of metal alloys. Chromium, tungsten, and vanadium had become scarce during the war, as the major sources for them were in enemy-occupied or enemy-sympathizing countries. Even after the war, when supplies were limited, the National Emergency (NE) steels policy was developed to preserve the scarce supplies. The shortage of landing craft (due more than to just available metals) was one of the considerations, among others, that postponed the date for the Normandy invasion of Europe from 1943 to 1944.

The naval boats, known by various acronyms such as LCVPs (landing craft, vehicle, personnel), or Higgins boats, were first used by the Marines for amphibious landings, beginning with the Guadalcanal campaign. On February 19, 1945, these boats set down on the volcanic sands of Iwo Jima, discharging sixty thousand men of the Fourth and Fifth Marine Divisions. Once the fighting commenced in earnest, the Third Marine Division, held in reserve on ships at sea, came ashore. Among them was one of Lockport's own, lieutenant and platoon leader Dominic Grossi, who was to receive the Navy's highest honor, the Navy Cross, posthumously for his heroism in leading his men in an assault on the enemy-held airfield. Steel, whether in armor plating for ships, landing craft, tanks, and other armaments, played a significant role in the destruction of the zealously defended Iwo, and other outposts of the Japanese empire.

James V. Forrestal, the Secretary of the Navy, acting for the President in awarding the Navy Cross to Grossi, had earlier addressed a crowd at Bethlehem, the nation's top war contractor. "This is a war of steel," Forrestal said. Indeed, steel was on every combatant's short list of immediate needs. Dominic Grossi and his father, Pat, were linked to that war effort by steel: Dominic, as a combat officer hitting the beach from a steel-armored boat to fight and die for his country; and his father on the home front participating as a Simonds steelworker. There were a number of young Lockport men who saw combat service: Cpl. Dan Rosati

(1919–1945) lost his life from wounds sustained in February, 1945; his brother, Al, lost his, a year earlier when his plane was shot down over Germany. A wooden plaque bearing Dan's name remains on a Simonds wall to this day. We were not related, but I remember his parents, Nick and Rose, who lived close by on Ontario Street. Other young men survived the ordeal and came home to be steelworkers or to begin other jobs.

Men who were working in defense plants, and who did not enlist in the armed services, or who were not qualified for one reason or another, made a contribution to the war effort; at least most felt that way. It was a most patriotic time in the country's history. My father, his uncles, my grandfather, and godfather were among those men at Simonds whose allegiances had transferred from Italy to the United States when the country went to war.

My father immigrated to the U.S. in 1933 and joined his father, Luigi, who was working at Simonds at that time. Luigi and his brothers, Francesco and Vincenzo, had first tried finding work in Brazil and Argentina, but they eventually came to America and Lockport between World Wars I and II and got jobs at Simonds. My great-uncles, Francesco and Vincenzo (aka Frank and James, or Big Jim at Simonds), were part of the steam hammer crew. My grandfather, Luigi, worked there as well. My father hadn't seen much of his father with all of his comings and goings to South and North America to look for work. He finally was able to join him as a seventeen-year-old in Lockport. He spent the first few months in America trying to learn English. He was fortunate to have an Italian-American school teacher at Charlotte Cross School in Lockport's west end, Miss Donatelli, later Mrs. Kenny, to teach him at a time when there was limited bilingual education for Italians or other immigrant groups. Several months after he arrived in Lockport, he was able to get hired on at Simonds part-time where his father was working on a more or less full-time basis. One evening in 1936, Luigi came home from work with what appeared to be a bad cold. My father told me it was "the grippe," which was a term used then for a respiratory illness that resembled influenza. His condition worsened over the next few days. He became too ill to work at the plant, eventually taking to his bed. The family doctor was called to the home and diagnosed pneumonia, a common complication of influenza. In those pre-antibiotic days, pneumonia, influenza, and tuberculosis were the leading causes of death in the U.S., as they were in much of the world. A patient's immune system either overcame those crises on its own or it didn't, and my grandfather was one of the unfortunates who didn't survive. He was a comparatively young man at fifty-three years of age.

When my grandfather passed away, he and my father were boarders at the DiGiovanni home on New York Street, around the corner from my future maternal grandfather DiPaolo's home on West Avenue. The neighborhood included several men from the Abruzzo village of Roccamorice. The grandfathers knew each other as young boys growing up there in the 1890s. Their immigration stories had different twists and turns in their paths to America, but they became reacquainted as Simonds steelworkers. My parents had many occasions to meet in the neighborhood at Saint Anthony's church services and social events held in the church hall. In 1937, a year after his father died, my father, who was working part-time at Simonds, married my mother, Marie, and they moved in with my grandparents at 353 West Avenue.

They lived in my maternal grandparent's home for the next few years. I was born there in 1940. During the days and weeks when my father was not called to work at the mill, he and my mother worked area farms picking whatever fruit and vegetables were ready for harvest. One of the retired workers who had worked with my dad years later, told me about the labor gang stories, and how my mother's great homemade bread got my father more work assignments than he would have gotten as a single guy.

The winds of war brought the demand for more steel, and by 1940 when I was born, my father had acquired a full-time job before becoming a U.S. citizen in 1942. It was during the war years that our family began its ascent into the middle class; but it would take another decade to get there. By the end of the war, the mill roared twenty-four hours a day, seven days a week, with the men frequently working double shifts; in the process, becoming men of steel.

One Sunday afternoon not long after the end of the war in 1945, our family was having dinner after Mass at the picnic table under the grape arbor at my grandparents' home not far from the plant. Afterwards, my father left the table to take a nap in the back seat of his old Plymouth that was parked in the driveway a few yards away at the side of the house. Not long after, while my mother and I remained talking with my grandparents under the arbor, the plant superintendent at the time appeared, inquiring after my father. My mother knew my father was tired from working double shifts for of the past six days and was reticent to speak. However, before she could come up with a deception, the super spotted my father's feet hanging out of the car window, walked over and woke him up. "Tony, I need you to come to work this afternoon." My father grudgingly got up and followed him back to the plant. That practice continued into the '50s. My brother recalled a phone call around 8:00 a.m. as he was getting ready to leave for school. It was Wally Jaynes, the personnel director. His son, Wally Jr., was my brother's classmate, and would eventually be employed at Simonds in production control. Wally Sr. wanted to speak to my father, who had just gone to bed after his 11–7 shift. When my brother told him that my father was sleeping, Jaynes responded, "That's okay, wake him up, I need to talk to him." My brother reluctantly put down the receiver and went upstairs to wake my father. When my father put the phone to his ear, he listened for a few moments then simply responded, "Okay." He got himself ready and went back to work. In those days, plant managers and their directors still retained sufficient power and influence. Few refused their bidding. John Coleman told me that his coworkers complained they spent more time at the plant with each other than they did with their wives and children. A separate life in the steel mill was a reality that was grudgingly accepted.

12
TAKING IT TO THE CLEANERS

When Michael finished shooting the old cafeteria with its sagging tray counter, dilapidated cabinetry and peeling paint dangling from the ceiling, we made our way over to the bar mills. If melting was one of the major aspects of the Simonds operation, rolling out the steel was the other. What was left of the bar mill was a series of grooved rollers in corroded housings, tarnished totems amidst the ruins, resting on a deck littered with gray granular debris and shards of metal. We stepped around the clutter, glancing up at the dusty steel joists that angled down from the ceiling. A multi-paneled window with broken panes in a checkerboard pattern focused light on the grooves of the first set of rollers we encountered. They were wider than the ones farther on, so judging from the Malcolm and Coleman diagrams, we were at the 16-inch bar mill. Rusting annealing furnaces stood adjacent to the mill, and the last set of rolls of the mill connected to a conveyor belt and metal ramp that looked like the collection point or hot bed for the newly rolled bars. We walked over to the 10-inch mill stand that was similar except for the narrower grooved rollers and the lack of a conveyor belt and collection ramp. Layers of dust covered racks of rollers in various dimensions beside each mill.

The ingots and slabs began to take shape on those mills, but before that, they had to be ground to provide smooth surfaces which would resist breakdowns as they passed through the rollers. When Simonds first began operations in 1910, they had what they called a chipping department. The ingots that came out of the molds were not perfectly smooth. They had surface imperfections, bumps, and other irregularities that had to be removed manually. Men using chisels bent over the ingots to chip away the bumps. In time, the chippers gave way to the grinders.

Like the Pickle House, there were many tales about the grinding operation. Euphemistically called the cleaning department by the company, it had nothing to do with maintenance, but was a battery of ten swing-frame grinders that were suspended from the ceiling. They were located near the rolling mills not far from where Michael and I stood, but we could find no trace of them. The originals had long frames with handlebars and abrasive wheels as large as tires.

Anthony D'Angelo, the son of legendary labor gang boss, Tony, was a crane man in the cleaning department. I asked him over the phone what the job was like. "Hot and dirty," he replied, pausing to clear his throat, "but the money was good." Anthony's job was to crane billets to and from the annealing furnace and deposit them on the grinding stands. The foreman would circle defects he spotted with yellow chalk, and then the grinders took over. The redness had faded, but the ingots still radiated heat. Their eyes shielded by goggles, the only other protection they had were handkerchiefs fashioned as masks, and heavy aprons. The men leaned into their apparatus, pressed, and swung the wheel across the surface of the ingot. A yellow and red shower of sparks and black granules cascaded to the side, resembling a fountain display at Outwater Park on the Fourth of July. One man's debris fell toward the grinder next to him and so on down the line to the wall and out the doorway, where its prevailing path was toward the commissary.

Grinding was hot, heavy, and nasty work. Many of the men were pretty strong; others who persevered soon got that way. There were good incentives based on output, so many stuck with it, including my grand-father's brother, Bartolomeo "Benny" DiPaolo. Bartolomeo immigrated to America in the mid-1950s. He was happy to have a good-paying job so that he could earn enough to finance a home and send for his Italian family. But it was definitely a job for younger men. As my great-uncle aged, he transitioned to lighter jobs and retired in the true cleaning, i.e., maintenance department.

Grinding, despite its incentives, didn't appeal to everyone. When Gordon Martin started work at Simonds in 1959, he told me he was assigned to the labor gang and that the cleaning department was among his first assignments. He apparently had the knack for grinding. Toward the end of the first week his older brother confided that the foreman told him, "Your brother is a very fast worker; he'd make a great grinder."

"After I heard that," Gordon told me, "I really slowed down. One week of that was enough for me."

Many of the grinders were African-Americans, which raised suspicions of racism in the mill. Department photographs in the 1940s and 1950s showed relatively few black workers, except in the grinding department, where a third of the men in a 1958 department photograph were black.

Discrimination had been a labor issue in the early years of the steel industry, where race and culture had raised its ugly head in the big steel mills of the South and Mid-Atlantic states before and during the union organizing period. African-Americans usually got the hardest and dirti-est jobs. Although not an overt issue in organizing the Simonds local, there were isolated instances of racism at the mill, as there were in the community. Still, the employment of black workers in an especially "dirty" department was a red flag: Why would anyone freely choose to work there?

After federal civil rights legislation established the Equal Employment Opportunity Commission (EEOC) in 1965, employment discrimination could put defense contracts at risk. Consent decrees required that discrimination against women and minorities be addressed. When an EEOC representative was sent to Simonds, he attempted to make an issue of the fact that nearly all of the black men worked as grinders. "Why do you work in this department?" he asked one of the black workers. The worker pulled a pay stub from his pocket and showed the representative, "That's why," he said. The fact was that although the work was physically demanding and they labored in a dirty, sometimes dangerous environment, the great incentive program made grinders among the highest-paid workers at Simonds. They liked the work, and the tonnage arrangement allowed them to work less than an eight-hour shift.

Black workers may have liked the work and the money they made, but they had limited opportunity to advance to supervisory positions in the shop or become union officials. In 1971, it became a civil rights issue. "We have a problem," wrote Orville Lingle, Simonds' civil rights committee chairman, to Alex Fuller, head of the USW Civil Rights Department in Pittsburgh. He explained that Simonds had added a number of new supervisory positions, including two foremen slots in the mill; however, no black worker was considered for either position despite having qualifications and seniority. The situation reminded me of one of Deborah Rudacille's interviews with a Bethlehem worker: "Blacks and whites got along good . . . only thing was, whites didn't want anyone promoting to their jobs."

The Simonds grievance procedure was the only mechanism to address the problem locally, but management was not responsive and the committee was warned to "stay out of it." Lingle's letter, which was signed by thirty-five black workers, cited Title 7 of the 1964 Civil Rights Act.

This federal act stated that employers could not discriminate due to sex, race, color, national origin, or religion. Fuller subsequently wrote to the District 4 director, asking him to investigate the matter. How the issue was resolved is not clear; the paper trail I was following came to an end. However, at least one of the men, Dave Woody, who had moved from the mill to the chem lab, became the recording secretary in the union and served the membership well.

Simonds had several types of grinders and abrasives. Some of these removed surface and deeper inclusions while others made cuts through bars of metal. I had a photograph of a worker using what I thought was a grinder, but when I showed it to John Coleman he identified it as a cutting-type wheel. "You can tell by the crotch protector," he said. Another machine was a large, long-armed semi-automatic device called a billet grinder that was operated by a man in a cab on rails. A tragic accident involving the billet grinder remains imprinted in the memory of many former workers more than thirty years later.

The victim, Henry Tomzak, was a young, recently married man who had been hired into the metallurgy lab. His job required that he enter the plant periodically to coordinate metallurgy lab and mill operations. He had a happy-go-lucky attitude and was well thought of by everyone he worked with. One mill hand recalled him often whistling as he walked. "When I heard 'Popeye the Sailor Man,' I knew he was coming." The young man was likely in that frame of mind while walking innocently back to the lab late in the day, in the alley next to the finishing operation. He usually clocked out by 5:00 p.m., but his duties kept him at the plant later that day.

On the other side of a corrugated-steel wall, the billet grinder had bored the large wheel beneath the surface of a steel bar. The grinder was not intended to be used as a cutting tool, but after day-shift management left, it was sometimes used in that fashion. Revolving at high speed deep in the bar, the wheel apparently overheated. A sharp fragment broke off, passed through a reinforced wall and flew across the room and through a window, striking the young man in the neck while he was en route to the lab. The blow staggered him and he fell to the ground. Blood was everywhere. Fellow workers applied pressure in an attempt to staunch the bleeding while the ambulance was summoned. He made it to the hospital and surgery was performed, but he died several days later. "It would have been better for his wife and parents if he had died then and there," Adrian Sherman, who attended his funeral, told me. "It was a freak accident. The angle of flight a moment sooner or later could have made a big difference." It was one of those "What are the odds?" moments with sad endings. The billet grinder, assuming the burden of guilt, fell into a depression and eventually took his own life.

A year after the accident a new cold-rolled cutting steel, HT1, was labeled in Tomzak's memory. The Industrial Revolution had altered the connotation of *accident,* and the steel companies, both large and small, provided fertile ground for its tragic manifestations.

Output—production and tonnage—was the priority. Health and safety was a secondary consideration in the steel industry until the passage of the Occupational Safety and Health Act (OSHA) in 1970. But even before that, as companies realized that accidents cost time, money and manpower, it would take time to instill a culture of safety. "We had no helmets, no mask, nothing," Tony Parete said, reminding me of my 1950s work gear. Safety began in earnest with written rules and procedure after safety legislation was passed and included personal protective equipment such as helmets, goggles, earplugs, shields, safety shoes, and gloves. Protective clothing in the melt shop included fireproof hoods, coats, and pants.

Historical accident statistics during one six-month period in 1910 at Sparrows Point listed ten fatalities, 304 "severe" accidents and 1,421 "minor" accidents. In the book *Forging America, The Story of Bethlehem Steel*, the Lackawanna plant's accident stats from 1905 to its closing in 1996 cites 650 workers dying in work-related accidents, mostly before World War II. Unsafe conditions continued long after fair wages and practices were addressed; but at Simonds, differences in the very nature of the labor conditions in the specialty steel mill precluded comparable accident data. I never thought my father was in any danger when he went to work, and I never considered I might get hurt when I clocked in that summer. Safety signs were posted around the plant, I remember, and tarnished vestiges remain, moldering away with everything else.

In time, recordkeeping of accidents and near misses, equipment maintenance, charts, graphs, and prominently placed signs indicating the number of days worked without an accident would become commonplace in steel mills and other industries. And there would be incentives, awards, and rewards for workers who practiced workplace safety. Safety improved over the years but had a long way to go. It still does. Serious industrial accidents and deaths occur almost daily in this country. During the post-war boom at Simonds, one of the most problematic mills in terms of safety and injury was what the workers referred to as the "snake pit," the 10-inch bar mill.

13
THE SNAKE PIT

The hot rolling operation at the bar mill was as fascinating as it was labor-intensive and risky. Accidents occurred in every rolling operation sooner or later. You had to know what you were doing because things happened fast.

On my breaks I would sometimes wander over and watch the men work. Now, standing next to the old mill housings, I thought back to when they were in active service. Most mills had a crew that included a heater, who managed the heating and softening of steel in the annealing furnaces. Steel had to be heated above its re-crystallization temperature, at least 1700 degrees Fahrenheit, so that it could be shaped and formed. Depending on the steel mill, the other jobs included: a heater's assistant, chargers and assistants, roughers, catchers, finishers, drag downs, hook ups, hot bed men, buggy or trolley men, and a screw operator. All in all, each rolling mill might have a dozen or more men on each shift.

So now, as I stood there, I gripped the tarnished screw handle connected to a circular gear that was integral to the housing of the skeletal remains of the mill. It didn't budge. I imagined how the screw operator adjusted the gap between rollers, using the lever to bring the billet close to a specified gauge. It reminded me of my mother adjusting her hand-operated pasta machine as she cranked out strands of spaghetti from billets of dough.

Bill DeCesare worked on the 10-inch mill for several years. His father, Tullio, a gentle giant of a man, was a good friend of my father and had worked with him in the '40s and '50s. Tullio was a heater; one of his son's early jobs on the mill was to hook up the steel from the furnace with chain-suspended tongs and maneuver it to the first set of rollers. The roughers, catchers, and finishers took over, using their tongs to grab snakes of hot steel shooting through the rollers. I marveled at how they snared the fiery rod, abruptly turned, and in one smooth motion swung it around their shoulders, and shoved the lead end back through an adjacent roller as the hot metal curled inches from their body. The 10-inch mill processed smaller, forged billets and slabs; rounds, and flats and diamonds, as they were called. The mill also rerolled some of the bars from the 16-inch mill. A 50-pound, 3-foot-long billet could end up 50 to 60 feet long with a diameter of one-quarter to one inch. When the mill was run at high speed, the bar would come through the rollers at 40 miles per hour.

The workers in the snake pit had to be on their toes. Most were quick with their tongs, catching and looping the snakes back through the rollers. But if one of them made a mistake and didn't catch it properly, there was no telling what that bar would do or where it would go. Some of the men had worked the mill for years. For them, catching and returning the bar became as routine as summer weather reports of triple-digit temperatures in Phoenix.

The steelworker's speed and timing with tongs was like a batter facing a major league pitcher with a 105-mph fastball. He couldn't wait to see the ball before he initiated his movement to swing, just as the mill worker couldn't wait to see the bar emerging from the roll and grab it with his tongs. He had to have the jaws of the tongs ajar, listen for the "click" and anticipate the bar's emergence. Like a batter, if he waited to see the ball, it would be by him for a strike, or he might foul it off into the crowd where something bad could happen. The analogy can only be taken so far. Of course the worker couldn't "wait for his pitch," or pass on a ball outside the strike zone. Just one strike or ball and you were out at the bar mill.

Less-experienced hands hoped and prayed something wouldn't go wrong; but of course, it sometimes did. A worker who failed to catch a fast-moving wire cable had it bounce around the floor and ricochet up to penetrate his upper leg or groin area. In the recollection of the workers who were there at the time, there are some differences in opinion regarding the exact location of the bodily penetration. One observer told me it passed through the soft tissue of the man's thigh, but others remembered it differently. The ricochet didn't kill him, others related, but his genital or GI tract required significant reconstruction.

Harold F. Kinsler told me about his rookie experience on the bar mill. He began his career at Simonds in 1961 in the labor gang, after graduating from Lockport High School in 1956 and a stint in the army. His father, Harold Sr., was a long-time employee at Simonds and a master roller on the 10-inch mill. "I don't want to work with my father," Harold told the personnel manager. But he was assigned to his father's mill despite his protest.

"Just do what I tell you, and everything will be fine," his father told him. He turned out to be right for the long haul. Harold Jr. successively performed all of the jobs on the 10-inch mill over time, became a crew boss, eventually replacing his father when he retired. Seventeen years later, he had become the supervisor of the entire rolling mill division, reporting directly to the plant superintendent. He eventually left Simonds to start up his own rolling operation.

Fortunately, an incident that occurred on the first week on the job had not discouraged him. They were running a particularly long strip, which required a fellow worker to catch the first loop and Kinsler to grab the second. Similar to a racing crew rowing together, each member fully in concert, confident in each other's skills, the mill worker had to rely on the guys he worked with, watch his back, and trust them to execute. When everyone was in sync, things went smooth as silk. But that day his fellow worker failed to catch the first loop, and the strip flew up and over Harold's head and down his back, searing through his clothing and inflicting a burn. Since he could still walk, he was told to go over to first aid and then come back to work after he had been patched up. "I'll go, but I don't know if I'm coming back," Kinsler said. Persevering past the first day or week at the steel mill was like studying anatomy during the first weeks of medical school: After the head and neck, it was downhill the rest of the way.

Few men with long careers at Simonds escaped injury, mostly leg burns. Facing the ovens and red-hot steel with face masks didn't always protect the worker, either. The plastic shield would heat up, and if it moved, brushing the nose or cheek, a trip to the first aid clinic might follow. In the large plate mills, radiant burns occurred when workers got too close to the 2000-plus degree metal. Contact burns were prevalent in the bar and sheet mills. Red-hot steel quickly burned through clothing and sliced open skin, leaving many with indelible scars. With small-caliber rods at high speed, the 10-inch mill was probably the riskiest mill in the plant; at one time that risk was reflected in wages. The foreman's salary on the 10-inch mill was higher than that of the 16-inch mill foreman.

Louie Koel worked on the 16-inch bar mill in the 1960s. Louie's story of coming to America in the '50s was much different than most of the immigrants I knew. He was born on the island of Saaremaa in the Baltic Sea off the coast of Estonia in the late 1930s. The Russians had taken control of the country and had sent his father and grandfather along with other men to Siberia to work in the mines. At the outset of WWII, the Germans invaded and occupied the Baltic states. Later in the war, when the Russians recovered and went on the offensive, the German officers gave the people of the Estonian town where Louie was living two options: stay and live with the uncertainty and consequences of Russian re-occupation, or evacuate with them to Germany. Louie's family chose the latter, and with his mother, two aunts, and cousins, they were shipped to Germany in cattle cars. Huddled together in the boxcar, they had no idea where they would end up. After intermediate stops, they ended up living in abandoned army barracks in the southern German village of Bad Reichenhall on the Austrian border across from Salzburg, where they remained for the duration of the war. In 1951, the family immigrated to America and settled in Schenectady, New York, where an older cousin was living. News that Harrison Radiator Division was hiring in response to the increasing demands for General Motors automobiles led the family to move to Lockport in 1953.

I met Louie while we were in high school. Lou was by far the biggest guy on the football team. He had aspirations to play college ball, but that didn't pan out, so he joined the military and was assigned to the Fourth Armored Division. In a strange twist of fate, his unit shipped out to Germany, where they were stationed on a military base that housed the old barracks where he had spent the last years of the war. "I felt right at home," he told me.

After his military service, he was hired at Simonds. He remembered his first day at work. Big and strong as he was, he told me he couldn't believe the noise and how difficult the job was. He hadn't expected the hard physical labor; he told me he was ready to quit after the first day. His story was similar to one Bill DeCesare told me about a nineteen-year-old who expressed a similar sentiment his first week on the 10-inch mill. "You've got to be crazy to work here," he told his foreman, who had informed him that he was going to have a hefty paycheck that first week. "Keep it, and take your wife to dinner. I'm not staying." New hires usually had a thirty-day trial period before they were permanently hired into a union job. Workers who couldn't cut it usually quit or were informed by management long before the thirty-day period expired.

Lewis Malcolm was Louie Koel's foreman on the 16-inch bar mill. He credits Malcolm and others at Simonds for training him and instilling a work ethic handed down to his generation, even though not everyone grasped it. The 16-inch bar mill, which processed larger and heavier ingots than the 10-inch mill and at slower speed, had four sets of rollers, each housing two grooved rolls. As in the 10-inch mill, the heater took a billet out of the furnace using tongs suspended on chains. His assistant moved the ingot to the first set rolls where it was "chained up," the phrase he used.

"What does chaining up entail?" I asked Lou.

"Each roll stand has a top and bottom roll. The roughers and catchers stand in the front and back of the roll stand. The billet is chained up to the roller and the first pass is made. Then on the back side, the hook-up man places a long-handled steel hook, suspended on a chain from an overhead girder under the bar and helps the catcher lift it. Then it's shoved through the top roll, and this process is repeated through the second and third roll stands."

Louie had had various jobs on the 16-inch mill, including hook-up man. He explained he had to bend over and peer through the rollers to judge where to place the hook under the emerging bar. Too soon or too late and the bar's thrust would cause the hook to fly off wildly. It was not only tricky, it was back-breaking bull work using one's back as a lever, and the men on the mill often tried to outdo each other lifting heavy loads. It was a "macho" thing before that term came into the general lexicon. Unknowingly, the work would take its toll. Steelworker gatherings in later years would find many complaining of bad backs. It was a Simonds legacy. Still another, the result of the constant pounding noise, was the hearing aid.

"The bar must cool off during these passes, right?" I asked. During all hot rolling operations, steel cools down and has to be reheated before it can be finally shaped and finished.

"Yeah, it starts out orange and ends up yellow, but you can imagine how hot that was."

I could. The 1921 Simonds catalog has a color chart illustrating the visual portion of the electromagnetic spectrum of hot steel in moderate diffused daylight according to its temperature, from maroon at temperatures below 1000 degrees Fahrenheit, through a range of dark to light cherry reds, oranges and yellows, to nearly white above 2200 degrees. In the old days, the temperature of hot steel was visually checked with an optical pyrometer. Standing close to metal in the higher temperature range, the worker didn't need an optical pyrometer to know what hot felt like. Two different workers in different contexts told me a tale that still seems dubious to me, of men putting ice packs (dry ice) in their pockets to protect their groins. Temperature is critically important in the many phases of high-grade tool steel manufacture, and the men intuitively sensed it was also vital to testicular function.

"What was next?" I asked. "You sent me a photo of a guy with a broom on a yellow-hot bar."

"Before that, the bar goes through another set of rolls, and then the edging set with two rolls to square it off. Then, before it was finished, a straw broom was swept over it to remove as much scale as possible before the finisher put it through that last set of rolls."

"I saw the steam coming off the bar in the photo you sent me. Didn't those brooms burn up?"

"We would go through thirty to forty brooms a shift. You'd use one broom per bar. Between us and the other mills we kept that broom factory in Lockport busy."

I remembered those steelworkers around the mill as Louie spoke to me, sturdy guys gripping tongs while in work clothes and skull caps, stained with sweat and streaked with dirt, an iconic image in shades of gray in the glowing light from the hot steel bar.

Louie continued, "Those sets of rollers were in tandem, and you had to work fast as a team. There was a rhythm to the work, and you had to know what was going on. A piece of steel maybe starts out four feet long and ends up as a 20-foot bar. There were different kinds of steel alloys, some softer, some harder, and you had to remember what you were working with. You had to know where and how to hook it and catch it and how to put it back in, otherwise you'd burn your legs up."

The broom or sweep off was an entry-level position on the various rolling mills. The other jobs given to newcomers were the drag down and stackers, whose job it was to grab and pile the finished product. Newcomers on these jobs were often a target for hazing. On the 16-inch bar mill, the drag down had to haul the bar down a trough and lift it onto a pile, a job that was nearly impossible when mischief makers added lime to the trough. The senior crew stood by smirking as they watched the newbie struggle with the bar down the sticky-surfaced trough.

Gordon Martin and Dave DeLang had similar rookie experiences working the 30-inch mill as sweep offs. They both recalled being hollered at for not being more aggressive with their brooms. The first heat had Dave wondering if he would survive the shift. "Sweep it off, sweep it off, you're not working fast enough," Dave was told. "Harder, whack it harder," Gordon was told. "You're not hitting it hard enough," the senior man yelled, forcing Dave to break one broom handle after another and opening himself to criticism from the boss for wasting a resource.

Mentoring a new man was an important part of a senior worker's role, even though it wasn't spelled out in his job description. Several steelworkers told me how appreciative they were of the help they received from their trainers and for the skill set they acquired at Simonds. But not everyone was helpful. "Some guys showed you little and got a kick out of seeing you struggle," Dave told me. "It was the same in all the departments," he said; "Just part of working at Simonds." The harassment was not always limited to the newly hired. Adding grease to someone's gloves or sandwich, locking someone in the confined space of a locker or toilet stall, steaming and switching labels on soup cans, and water fights were among the daily pranks. There were plenty of jokesters and some real characters under the roof.

One guy frequently walked around the plant on breaks borrowing money but never paying it back, I was told. He made a practice of asking for change, never a dollar or more. "People would remember that you owed them a dollar," he said, "but no one remembers 75 cents." I was told he made several dollars a day. He may have been relying on the kindness of lenders: *And if you lend to those from whom you expect repayment what credit is that to you?"* (Luke 6:34).

The culture of card playing on work breaks spilled over after work into more card playing and other gambling venues. My father was one of the steelworkers who enjoyed the horse racing at Fort Erie and Batavia Downs. I was too young to accompany him in the '50s, but later in my college years I went with him on couple of occasions to watch and bet the harness racing at Batavia. Some men loved to brag about their winnings at the track. One steelworker in particular would bet every horse in a race to ensure he could show a winning ticket. An older retired worker told me a story, which was subsequently confirmed by others, about a school teacher in Lockport who worked the graveyard shift at Simonds and then left to teach for the day. When asked why he worked that hard, he responded, "I like to gamble." In retirement, many steelworkers would continue to gather in homes and bars with fellow retirees to play cards on a weekly, if not more frequent, basis. My grandfather was one of them, often playing cards down the street at Luigi Giansante's, the father-in-law of future steelworker, grocer, and Lockport alderman and mayor, Tom Rotondo.

■ ■ ■

And so we continued our journey. Leaving the bar mill area, Michael and I walked south in our path of discovery and came across a long-handled rusting steel carriage with two wheels propped on end against a wall. "There's a buggy!" I exclaimed. Pushing and pulling that rolling stand around had been one of my jobs when I had been assigned from the labor gang to the cogging mill. It had a pair of heavy steel rollers through which billets were processed for subsequent operations. I guessed we were standing close to the sheet mill area, which was the ultimate goal of my visit. The cogging mill received the billets after the hammer, or later, the press worked the ingots and prepared them for subsequent sheet mill operations.

As the "buggy boy" I worked with the heater and rollers. When the steel had been sufficiently heated, the furnace door was pulled open and the heater used a long-handled hook to yank out a red-hot, 2200 degree Fahrenheit billet onto the surface of the buggy that I had wheeled up to the furnace's surface level. The first time the blast of heat hit my face, I turned away and almost lost my grip on the buggy handle. Imagine opening the oven to check the Thanksgiving Day turkey, but quadruple the intensity. I would have dropped the ingot and dangled in the air had the heater's assistant not steadied me and the buggy. Regaining control, I wheeled the ingot over to the rougher, who pulled the ingot off the buggy stand and shoved it into a set of rollers that resembled an old-fashioned wringer washing machine. As the roller and his catcher shoved the steel back and forth to one another through the rollers, it gradually lengthened, cooled, and the pounding intensity diminished. At the appropriate moment, I was signaled to bring my buggy to the rollers and the elongated ingot was slid onto the carriage. The slab I returned to the furnace for reheating was much cooler. Over time, the rerolling produced a sheet that gradually crept up the handle of the buggy until its red-hot edge was almost a foot from my gloved hand. Facing the searing, red-hot sheet stung my face, so I turned my back to it. Gripping the buggy handle, I lifted my shoulders and shirt collar as I scrunched my neck to shield it from the intense heat and wheeled the sheet from the furnace for the return trip to the rollers.

A retired steelworker who had that buggy boy job when he was a new hire told me how one of the mill hands purposely and repeatedly dragged the hot sheet to within a few inches of his gloved hands. Finally, he had had enough and intentionally dropped the handle of the buggy, sending the sheet of steel sliding onto the floor close to the heater at the furnace. The angered heater's short but spectacular response ended that prank, if not forever, at least for that worker that day.

In no time it seemed, I became sweat-soaked and thirsty. The cogging mill was my first hot and hard job, and it came on a hot and humid June day in Lockport. I worked up a ravenous appetite, and I looked forward to a break when I could sit and have a drink and a sandwich. I never paid attention to what happened to those long plates of steel after I had delivered them on my buggy for the last pass. I discovered later that there was another crew of men that dragged them to the rugged block out shear located in proximity to the cogging, 30-inch and sheet mills. There the still-hot and heavy steel plates were lifted onto a table, pushed into the shears and cut into blanks that would be further worked on the sheet mills. The block out shear was another heavy-lifting job. In terms of work flow and supervision, the block out shear, the cogging mill, and 30-inch mill was organized as a single department. A 1958 photograph of the group shows thirty-six beefy men, most of them west-enders, one of whom was my father's cousin, Joseph (Danny) D'Angelo, sometimes called "Muscles" around Simonds. He was born and grew up in a tough New York City lower east side Italian neighborhood and later moved to Lockport, before serving in World War II and getting a job at Simonds. Although he could be aggressive (one of his sons, Robby, became a boxer) he also had a gentle side. He told me one day at work that Simonds was "no place for you. Be happy you're in college." He eventually took his own advice, left the mill and opened the Sugar Shack Restaurant in Lockport's west end.

At the end of the day over the first couple of weeks on the cogging mill, I dragged myself home and went straight to my bedroom for a nap before dinner. After dinner, I was still tired and had no desire to play in the summer softball beer league, but I was proud of meeting the physical challenge of the job. When I started working at the plant in June, I brought four sandwiches and assorted fruit and cookies in my lunch pail and devoured everything. By the end of August, I had adapted to the work, reducing my intake to a reasonable single sandwich, and I played ball after dinner.

My status in the labor gang had me moving around the plant week to week, sometimes day to day. I would have preferred a steady job in the melt shop or bar mill, but realized the likelihood of a temporary worker being assigned to those departments was slim. I looked forward to handling the buggy as much as possible, finding it much preferable to the sheet mill finishing department, where I repetitively stacked steel circles as an assistant, sitting behind an operator who manually punched out the circles from sheets of cold roll steel. Eventually, years after I left, that operation became fully automated with entirely new machinery. I had one week left to work before going back to college to begin my second year when I was given the opportunity to work the third shift on the sheet mill, rolling steel with my father.

14
THE SHEET MILL—STARTING OVER

" I think the sheet mill would have been over there," I said to Michael, pointing to an old furnace next to a raised platform, trying to recall the sheet mill department now more than fifty years later. I knew that the actual sheet mill, the rollers and housing that my father and I had worked on had been auctioned off during the bankruptcy sale. "Your grandfather spent a lot of years there and I'd like to walk there again." In an earlier winter visit, Michael had taken a shot of the annealing furnace and had given me a print. It was an artistic composition of a snow-dusted, rusting furnace and ground with a stack of long-handled tools illuminated by the natural light from missing overhead roof panels.

Shift work: My father had worked some erratic ones in the old days when it was common to change shifts (7–3; 3–11; 11–7) each week on the swing-shift schedule. Deborah Rudacille's memoir, *Roots of Steel*, is set in Dundalk, a blue-collar steel mill town outside the massive Bethlehem's Sparrows Point steelworks. She writes that rotating shift work was tough on the men and family life—especially during the early days when blast furnaces were only shut down once a year. Back then, men were required to work twelve-hour shifts, six days a week, and then a twenty-four hour shift on Sunday before getting a day off. It was a tough, repetitive life of work, eat, and sleep. Rudacille quotes a saying during those 24/7 years at Bethlehem: "There are no Mondays, just yesterdays and tomorrows."

Progress in work practices had come with newer furnace technology and the union movement. When the entire crew swung together on five-day weeks, they would have weekends off. I certainly had it better at Simonds working the 7–3 shift all summer long. My father was a foreman on the 11–7 shift then, so I had never worked with him. I wasn't thrilled about the 11–7 shift, but there was no other way to work with my father and his crew. I agreed when I was asked if I wanted the opportunity my last week on the job. I thought I had adapted to the labor by the end of summer, but I didn't know what the sheet mill job would be like. My father didn't tell me what to expect. Perhaps he thought I knew. The shift change had me feeling like I was starting my job at Simonds all over again.

My dad was always supportive, but in a quiet way. He was man of few words. His eyes did the talking, and they could transmit his disappointment, or pierce me with warnings. He grew up largely without the guidance of his father, who traveled back and forth from Italy to Argentina and the United States looking for work. My dad didn't advise or encourage; those nurturing roles fell to my mother. He was the provider; and for me, this short, stocky man was my model for a solid work ethic. And he was the ultimate disciplinarian. "Wait till your father gets home," my mother would say when she felt her spanking was insufficient to my misdeed. One swat of his calloused hands on my behind would lift me off the ground, and have me squealing, "I'm sorry!" through a cascade of tears. In that respect, he probably wasn't so different from the other neighborhood fathers of that immigrant era, men who kept their sons at an emotional arm's length while providing discipline and authority. On the job, he joined in the comradely behavior typical of a Simonds mill crew. He could be entertaining and engaging; but he could become upset when his authority was questioned or ignored. He was a tough man to work for, I was told. A hard worker of the "old school," he demanded the same from his crew. A worker told me how angry my father got one day when his crew ignored his call to stop playing cards and return to their jobs after a roll change. My father got up on the table scowling, jumped up and down, sending the tabled cards flying, yelling, "Back to work!" They quickly left the table and resumed their positions. No one spoke a word for the remainder of the shift.

He could also be stubbornly persistent in his way of doing things, like a one-way, "my way" traffic sign. There was an incident, I was told, when that attitude got him into a predicament from which he had be rescued. It came about when the sheet mill was down for repairs and he was sent to temporarily oversee operations on the 90-inch mill. His insistence on rolling the billet his way ran counter to how it had usually been done. One of the crew, an ex-Marine, a man twenty years younger, a foot taller and fifty pounds heavier, took offence when my father repeatedly ignored his suggestions. At the change of shift he waited to settle things with my father in the parking lot. The plant superintendent had been informed about the situation and wouldn't let my father leave. His antagonist was persistent and laid siege. My father was forced to wait it out in the office for about three hours until the angry worker finally drove off.

I knew nothing of this history when I awoke on the Monday morning of my last week at Simonds to a hot and humid day, happy that I wouldn't have to face the stifling heat inside the plant that day. I busied myself with getting things ready to go back to college while mentally gearing up to working with my father in what I hoped would be the cooler temperatures of the evening. After sunset, my father and I sat on the front steps of the house as we sometimes did on hot nights, hoping that the air would cool down before going to bed. But bedtime that night would be postponed. At 10:30 P.M., I slid into the passenger seat of my father's 1957 Chrysler Saratoga, with its push-button drive, two-toned coral and white paint scheme, plus all the chrome and big fins of the Chrysler Corporation's "forward look." There wasn't another like it in town. A middle-class job and a frugal lifestyle back then could buy that car; our house was paid for and I was in college.

I didn't know what to expect as I clocked in for the graveyard shift. As I walked from the gate, I passed the 90-by-30-inch mill to reach the 26-by-42-and 24-by-54-inch sheet mills. (The first number refers to the roll diameter, the second to its width. Workers usually referred to one number or the other, or simply "the sheet mill.") I was struck with how different the plant looked during the incandescent hours. With fewer workers and departments operating, it was much quieter. By the start of the 11:00 P.M. shift, the temperature in the plant had become reasonably comfortable.

Stepping onto the sheet mill platform, I could look into the mill's finishing department, where I had worked many day shifts. The large space with high ceilings and windows that was brightly lit by day now had little ambient light, a space filled with shadows and silhouettes of steel piles and equipment. There was no one working the annealing furnaces, shears, and punch machine. I didn't know what my job would entail that night, nor that I wasn't prepared. I was happy not have to stack circles of punched-out steel in the finishing department, but it turned out that I would be stacking the sheets themselves.

The sheet mill was similar to the bar mill, except for the smooth rollers in the housings in tandem on a raised steel platform a few feet from the furnace. The operation began with the heater pulling the hot blank, as they called it, from the red glowing furnace and placing it on a rolling stand. At the first set of rollers, the blank was shoved in with a startlingly loud bang. Then rollers and catchers, first roughers and then finishers, fed and returned the steel sheet to one another while the screw operator levered the rollers closer to the sheet's specified gauge. Just as in the bar mill, I was amazed at the ease with which the men turned, flipped and rolled sheet after sheet, their voices muted by the ambient sound of steel on steel. The rolling routine had become second nature to them. They knew their jobs and each other so well, they didn't need to articulate. Visual cues and body language was the mode of communication.

The master roller on the second set of rolls finished the sheet based on his "eye balling" experience before confirming it with a pair of calipers. My father was the finisher on that shift and had to remove his gloves to use the calipers. The sheet was still hot, so the caliper was a long-handled device. I was told it was often recalibrated with a set of standards to meet the tight tolerances required for quality control. When the heating, rolling, cooling, reheating, and rerolling sequence was completed and the sheet was slid over to me, it was my job as the stacker (or in the steelmaker's parlance, the singles boy), to lift the sheet with my tongs and create a pile that would eventually be moved to finishing and cold roll departments. The term "singles boy" derived from sheet mill operations where the job required separating and stacking multiple sheets that had been rolled together.

My father had learned the finishing roller's job like his predecessors, by first doing all of the other jobs on the sheet mill over a period of years: jobs learned more by doing than watching and listening. He had been a heater's assistant, rougher and finisher, and was mentored by Bill Oakes and Johnny Apolito, who had preceded him on the job. Manipulating the heavy sheets through rollers with steel tongs took physical strength, particularly at the front end; and over time he, like others, had developed Popeye-like forearms, which I wish I had when I was an aspiring baseball player. He had also developed thick, calloused hands. I remember him picking at calluses you could burn with a cigarette and not even feel. It took time and experience to become a master roller, but he finally got his chance and I was proud of him.

Around 1:00 A.M., after two hours without a break, I was struggling to keep up with the work. At first, flipping the sheets of steel with tongs to gain momentum for the lift onto the stack was not difficult. These sheets were of a relatively thin gauge designated for hacksaw blades, but they were unwieldy. As the stack grew taller, the job required more effort. The stacks I built were destined for the sheet mill finishing area. At that point, the sheets were ready for inspection and shipping, or, depending on the order, finished up by one or more processes such as pickling, annealing, shearing, and flattening.

Cold rolling was another treatment after hot rolling at some sheet mills. At Simonds, most of the cold rolling was performed on steel from the band mill. I had heard the term "cold roll" often, but the process was an unknown to me at the time. The operation was in a separate building, and I had never been sent there to work.

The heart of the cold roll department was the Z-mill (Sandzimir) machines. The term "cold roll" is somewhat misleading. The rolling is done at room temperature, which is well below the re-crystallization temperature of steel. The re-crystallization temperature, above or below a certain temperature for each alloy, distinguishes hot rolling from cold rolling. Originally developed in the early 1930s, the Z-mills had a uniquely clustered roll configuration. The machine's operator was able to roll strips of stainless steel cut from sheets on a metal bandsaw specially built for Simonds. The Z-mill transmitted force, reducing the thickness of the steel to very narrow gauges. At the same time, the pressure altered the sheet's grain structure, imparting strength and a hard, shiny surface. Finally, the machine spooled the long steel strip into large heavy coils. Because many customers wanted their steel less hard and more ductile, crane operators then swooped in to pick up the coils and transferred them to one of a dozen or more Bell annealing furnaces, where the heat altered the crystalline structure of the metal and made it more formable. Some steel orders required annealing before cold rolling. After annealing, the coils were slowly cooled; some were sent to the finishing department where they were punched into rounds to make home use-type circular saw blades. Most of the coils, however, were shipped to Fitchburg and to other customers who made hacksaw, bandsaw, and other types of blades. In 1972, new equipment was added to enhance the company's ability to roll and shape edge wire for bimetal band saws.

Some of the workers on the hot rolling mills ribbed fellow workers in the cold roll department for working in what they called the "old folk's home," a term that they indiscriminately applied to finishing and inspection work in general. That was because the work was performed at ambient temperature, it was cleaner, and machines did most of the heavy work. Joe Murphy worked cold rolling for a number of years, but when he started out at Simonds, he began in the hot rolling mills, where one of the seasoned men advised him to save himself a lot of grief by looking for an opening in the cold roll department.

But danger lurked in all steelmaking departments, and the cold roll was no exception. One of Simonds' fatal accidents occurred on the Z-mill. A worker, who was wiping down a roller while it was still in motion, caught the edge of his shirtsleeve in the roll. He was dragged in and crushed. I'm sure it was a horrible sight; it was a horrible thought. But as I stacked sheet after sheet and made pile after pile on the sheet mill that night, all I was thinking was *When are they going to call a break?*

I didn't want to disappoint my father by lying down on the job, but I was worried that I was not going to make it. Even though I had the easiest job on the sheet mill, I felt beat, as if it were my first week on the job. My father might not have said anything to me, but at the time

I didn't know he had a reputation for being a demanding boss. In any case, I didn't want to embarrass him in front of his crew. I didn't want to admit that I didn't have the stamina to finish the job. I was ready to "throw in the tongs." Usually on the rolling mills, the breaks came reasonably often. Between the heats, men would relax, drink, and play cards, but that night the first break was a long time coming.

The work suddenly stopped at 2:00 A.M. The crew retired to their rest area, where they sat down at a table and chairs to snack, drink, and play cards. I lay down on a bench in exhausted anticipation of work resuming in fifteen minutes. I closed my eyes and nodded off. When I opened my eyes and sat up, I saw that my father and the crew were playing cards. We never returned to work that night. The shift was responsible for turning out a specific tonnage, we had reached that mark in two hours of steady work.

Tonnage was a wage incentive plan in various departments at Simonds, and it behooved everyone on a crew to work together as a team to maximize wages. In the early union-organizing days at Bethlehem, this concept had not been popular, and it was sticking point in early union business. Workers at Bethlehem felt they paid the price in lost wages for circumstances they couldn't control: downtime primarily caused by management. However, workers at Simonds seemed satisfied with the arrangement. I just wished my father would have told me the routine, but I had now been prepared for the next four nights. While the crew played cards, ate and drank (not beer, as I remember), for the remainder of the night, I slept until dawn. As the morning sun lit the plant, I rose, clocked out at 7:00 a.m., and went home to bed again.

That week on the sheet mill was my last week of work at Simonds—the last time I had been in the plant. I had wanted to capture that experience for posterity, to retain the memory of working with my father, so on the last night I brought my camera to work and shot a roll of photos of the crew at work, including a posed shot of my father and me holding our tongs in front of an annealing furnace. A week later when I picked up the photo package at the drug store, I discovered that none of the photos had turned out. My Kodak camera, a folding bellows-style model with a flash attachment, had never failed me before, and I was quite dismayed. I thought that I needed snapshots of the experience, otherwise it didn't happen. But it did, of course, and I have relied on a pathologist's visual memory to fill in details on these printed pages.

Our family's photo album had a single black-and-white snapshot of my father posing with six of his fellow workers on a break in front of the sheet mill, but no retired workers I spoke with had snapshots of themselves or coworkers at work in the plant. It may have been against company rules, as it was at Bethlehem. The only photos I was able to find of men at work at Simonds had been taken by executives or commercial photographers who had been hired to document various operations and processes for marketing brochures and catalogs. The company also took group photos of men in their various departments on two separate occasions: 1941 and 1958. Reprints of some of these appeared from time to time in the *Union-Sun & Journal*. Some of the work scenes could be reconstructed from steelworkers' oral histories; however, I found faded memories were sometimes a hindrance and a frustration for the men trying to recall events decades earlier, and photos would have helped. Luckily, newspaper microfiche articles with photographs, along with labor union archives I had found, could refresh their memories, occasionally bringing startled expressions and a sense of wonderment. But there was one remembrance that none of the workers was able to recall with any clarity, and that was the history of when and how their local union was formed and who their first officers were. I found this history especially important, because it explained an essential element of the steelworker persona, how their work was classified, performed and evaluated, how the jobs were taken for granted, why the loss of those jobs was so devastating, and why they grieved those losses for as long as they did.

15
UNION BUSINESS

It was mid-morning now, and it would soon be time to leave the building. We had escaped notice thus far, and it made no sense to push our luck. But I couldn't help thinking of how my life had evolved after Simonds, and how my father's and his fellow workers' days ended in the place where Michael and I were now standing.

As late as the 1950s, many kids followed their fathers into the steel mill. It was, for some, a family affair. Going to work in the mill was a good job. The salaries for workers with seniority allowed them to buy a home, an automobile, even send their children to college—in short, to achieve the American dream. The one refrain I kept hearing was "but the money is good."

Recent labor trends favor the college-educated worker's earning power. However, union wages once allowed workers without college degrees to live a comfortable life.

While I was still in elementary school I thought I would be a steelworker, but my mother disavowed me fairly early of that notion. "You will not be a steelworker," she said emphatically, "you're going to college." By middle school, I had put thoughts of the steel mill behind me. A lengthy stint in the hospital had introduced me to the medical laboratory, while daily contact with Dr. H. Braden Fitzgerald would become an inspiration for my career in medicine.

I had a year of college under my belt by the summer of '59, but I had also extended the Rosati lineage at Simonds for another generation. My son would not add a fourth generation, but I was happy that he had had an interest in the family history and had spent time under the Simonds roofs with his camera, documenting its abandonment and decay.

I went back to college in September, and my father went on to another twenty years in the mill. Some of my high school classmates, students of earlier and subsequent graduating classes, friends and cousins, joined him at the plant, finding jobs that paid well, had good benefits, and a pension plan they could look forward to in retirement. No one could have imagined that the steelworker would become discarded and abandoned like the buildings in which they once worked.

Before the steelworkers' union was established, steelworkers had to keep working through illnesses, until they suffered a disabling injury, or dropped dead. Once Simonds' family expanded their business in the late 1800s, they offered no health, accident, or retirement package to speak of until 1926.

That year, Simonds initiated the Simonds Saw and Steel Aid and Benefit Association. The articles of agreement called for open membership for all workers, and established a board of directors elected from its melting, rolling, shearing, labor, maintenance, and office departments. Monthly assessments at various levels from $.35 to $1.00, provided weekly benefits from $4.00 to $15.00. The death benefit was $100.00. The average cost of a funeral in upstate New York then was about $200.00, according to Anthony Farone, who checked the records of his four-generation mortuary business in Syracuse, New York. At a time when a steelworker earned less than a thousand dollars a year, the death benefit would have softened the blow for a steelworker's family.

The Aid and Benefit Association was an example of the state of benevolence in the industry when John L. Lewis, president of the

United Mine Workers, led a faction out of the American Federation of Labor (AFL) in 1935 and into a new organization, the Committee for Industrial Organization (CIO). Samuel Gompers's AFL, which he had formed in 1888, included his strongest union, the Amalgamated Association of Iron, Steel and Tin Workers, which had been steadily losing power and support in steel mills around the country. The union discriminated against unskilled workers, especially blacks and immigrants, and offered minimal support to workers on strikes, including the famous 1892 strike at Carnegie's Homestead plant. The steelworkers had refused a pay cut that year and were denied bargaining rights. Steel executives at the time believed in using any means necessary to keep workers from forming a union. While Carnegie was away on an extended vacation in Europe, his managing partner, Henry Clay Frick, an experienced strikebreaker, played a behind-the-scenes role in a strike that ended after a bloody gun battle between workers and Pinkerton guards. The Homestead strike was a game-changer, a clear victory for management, and eventually propelled the industry to new production heights. When the plant reopened, it was a non-union shop. Sixty years would pass before a national workers' union would emerge to grant workers bargaining rights for better wages, benefits and working conditions.

The steel industry's first decades of the twentieth century witnessed testy labor disputes and violence in steel towns such as Chicago, Youngstown, and Johnstown, and strikes in the coal and railroad industries. The labor problems occurred in a milieu of anti-capitalistic sentiment, anarchistic terror plots, and government overreach with raids and deportations that dimmed the glittering excess of the Roaring Twenties. The Wall Street Crash of 1929 closed out the era and ushered in the Great Depression, which led to the national mediation boards and commissions of President Franklin Roosevelt's New Deal's strategy for recovery.

In June 1933, Congress passed and Roosevelt signed the National Industrial Recovery Act (NIRA, aka the Wagner Act). The National Recovery Administration (NRA) was an essential element that authorized the president to institute a number of labor practices to assist the nation's economic recovery. It began its work with much nationwide excitement fueled by a massive public relations effort. The largest parade in New York City's history took place that summer, while in Lockport a more modest parade of industrial workers included Simonds employees. Among the NRA's various provisions was section 7(a), which recognized workers' rights to organize unions and bargain collectively. Not everyone was happy with the NRA's work. Business leaders, workers and union organizers were displeased with certain aspects and outcomes. Two years later, The Supreme Court ruled the act unconstitutional. Fortunately for the workers, subsequent legislation incorporated the labor provisions of section 7(a).

The CIO's intent was to organize the steelworking industry. Lewis' aim was inclusive, but he especially sought to improve the plight of the unskilled immigrant and black workers. He pushed for $5.00 daily wages, a 40-hour workweek and time-and-a-half overtime. My grandfather, Antonio, often praised FDR and the CIO in conversations about his work at Simonds. My recollection was that he was a strong union supporter; he proudly wore his CIO pin on his suit jacket lapel. I remember him often walking downtown to the CIO office on Main Street for some reason or other, perhaps to pay his dues.

The CIO had a turbulent history and might never have succeeded if not for the strong-willed effort of Lewis who, with his thundering presence, bushy eyebrows, scowl, and dogged perseverance established the Steel Workers Organizing Committee (SWOC) in 1936. Lewis put a highly competent aide, Phillip Murray, in charge of organizing. While Lewis and Murray pushed from the outside, workers pulled from the inside. The big steel companies had tried to counter early unionization activities with employee representation plans, the so-called company-sponsored unions. The NRA did not actively prohibit these unions. Lewis considered these company plans shams: "pious pretexts," he called them. They were drawn up by company lawyers and could not be modified without company approval. The company appointed the officials and influenced the assignment of production managers. It's likely that a similar illusion of representation existed at Simonds in the 1930s, policies clearly outside the intent of the Wagner Act.

But in 1937, a breakthrough occurred when a subsidiary of United States Steel, Carnegie-Illinois, unexpectedly signed a contract with SWOC. That year, Gifford K. Simonds announced the establishment of a $5.00-per-day minimum wage for Fitchburg workers (which may have been more than Lockport workers were receiving) in an attempt to blunt the appeal of unionization. Bethlehem, and the "Little Steel" companies represented by Republic, Inland, Youngstown Sheet and others, continued to advocate for their company unions. By the time of the Pearl Harbor attack on December 7th, 1941, the "Little Steel" companies had caved and signed contracts with SWOC, effectively ending the company union system. Bethlehem followed a short time later. On May 22, 1942, in Cleveland, Ohio, SWOC became the United Steel Workers of America (USW) with Phillip Murray as its first president.

The USW set about to establish local union chapters within a system of districts around the country as the steel industry ramped up production for the war. Pleas to *"Buy War Bonds"* appeared on flyers rallying steelworkers to *"Join the Union."* Under Murray's presidency and David MacDonald's financial guidance, the USW became a well-run, organized union whose aim was to make steelworkers the best-paid factory workers in the country. The national office controlled the union purse strings. They collected dues from the local chapters and distributed funds and directives back to them.

The task of organizing the steelworkers in Lockport fell to Joseph Maloney, Director of District 4, headquartered in Buffalo; and his representative, George Male, who coordinated the Lockport initiative. In October 1942, Male began communicating with the Lockport steelworkers, who were meeting at 373 West Avenue. His letters addressed to United Steelworkers, CIO, at 373 West Avenue, was the address of the Granchelli family business, a bar-restaurant and dance hall attached to their home at 371 West Avenue. The bar was accessible through a closet in the home. The hall could accommodate several hundred people for dances and other events, so it made a convenient venue for gathering steelworkers, many who lived in the neighborhood. The bar was named "Red's," after the proprietor, Joseph Granchelli, who the Italians referred to as *"Il Rouche,"* (redhead, in the Italian Abruzzo dialect). Two other Italian-owned bars in the area, the Rex Grill across the street, and the Red Robin, two doors up on the corner of Ohio Street, had much smaller meeting venues. I grew up in my grandfather's home three doors down at 353 West Avenue. By the time I became aware of Mr. Granchelli, he had closed the bar, remodeled, and expanded it for his ever-expanding extended family, and had gone to work at Simonds. Copies of those early communications to the West Avenue address have been preserved in the labor archives at the Penn State University Library Special Collections. They provide an insight into early unionizing activity in Lockport, although many information gaps leave the record incomplete.

Male's letters addressed the USW/CIO's desire to represent Simonds' workers in securing a contract with better wages, seniority rights, paid vacations, a grievance procedure and good working conditions. He pointed out the USW's signed contracts with over 1,200 companies representing 660,000 workers, and he drew their attention to the working conditions that employees at Harrison Radiator enjoyed under their CIO union contract. A USW application form for individual membership was included in one of the letters, as was a notice of a meeting to be held on November 1st at the Granchelli West Avenue address, referred to then as Steelworker Hall.

On December 6, 1942, the *Organizing Report of a New Local Union* listed a slate of officers which was submitted via the district office to David MacDonald, secretary-treasurer of the USW in Pittsburgh:

William Bryant, president
Alfred Golden, vice president
Elmer Gagliardi, recording secretary
Arthur Horton, financial secretary
John Sheehan, treasurer
George Fingerlow, guide
Clifford Lake, guard
Augie Sansone, guard

There were about 700 employees at the time, with 93 becoming members of the new Local Union 2857.

January 1943 saw considerable correspondence between the District office in Buffalo and the steelworkers in Lockport regarding informational meetings held for the rank and file on the 10th and 24th at the CIO union hall at 29 Main Street. *"Attend a Mass Meeting. Ask questions and hear good speakers,"* the posters read. *"We will have beer on tap."* The District enlisted the help and cooperation of people in the community, such as Antonio Valentino, the proprietor of the Rex Grill, along with Red's and the Red Robin, all popular Simonds workers' watering holes. William Hilger of the Lockport United Autoworkers union was also asked to help. The steelworkers were urged to join fellow workers in unions in Lockport and at the Allegheny Ludlum steel mills in Troy, New York, as well as the Sanderson Division of Crucible Steel in Syracuse. Workers in those mills were doing the same type of work at better wages, with eight-hour workdays, paid lunch periods, job protection, and other inducements. During that month, a conference between the officers, who held the provisional union charter, and Simonds management, was requested for the purposes of beginning bargaining—however, Allen Potts, the Simonds manager, refused, unless and until the union was officially certified to represent the workers by the National Labor Relations Board (NLRB).

The NLRB met in Buffalo on February 9th to hear arguments based on the petition of Local 2857 as the appropriate agent for collective bargaining. The steelworkers were represented by District 4 Director, Joseph Maloney. Simonds was represented by Lockport attorney, J. Carl Fogle. On March 1st the NLRB handed down a decision directing a special election by secret ballot to determine whether or not the approximately 580 employees at Simonds Saw and Steel wished to be represented by Local 2857. The *Union-Sun & Journal* on March 9th ran a half-page notice: "A Message to Simonds Saw & Steelworkers from Five Thousand Lockport CIO Workers," urging every Simonds worker to cast a ballot to join them in the fraternity of CIO unions. Male's final letter before the voting indicated that polling booths would be set up at the north end of the cold rolling department. The NLRB election was held on Friday, March 11th, the day Local 2857 was born. The official announcement came the next day in the *Union-Sun & Journal*: "Union Wins at Simonds." The report indicated that 383 out of 526 votes were cast in favor of unionization. All but 33 eligible voters cast ballots; 133 voted no union, a few ballots were challenged, four were void, and one was blank.

With victory declared, celebratory telegrams and letters of appreciation were sent and received in Lockport, Buffalo, and Pittsburgh. The work of electing a new slate of officers and beginning contract negotiations got underway later in the month, following a letter from George Male to Allen Potts to name a time and place to begin the process.

Who the first officers were that the rank and file elected following the submission of the initial organizing slate is not clear. The monthly union meeting records from that period have not survived, and the official ballots in the labor archives were not dated. Presidential nominees in the early '40s included William Bryant, Lawrence Rinaldo, Edward Cook, and Carlton Reid. Lawrence Rinaldo, who doubled as a steelworker and west-end tax preparer, served one term as union president and several terms as its financial secretary and grievance committee man. These and the other elected positions of vice president, recording and financial secretaries, treasurer, guides, guards, trustees and grievance committee men were elected by the Simonds rank and file who also elected stewards in their departments. Among the many West Avenue names I recognized on those ballot slates were my grandfather, Antonio, and his cousin Louis, who were nominees for guards, as well as several family friends.

These men worked their regular jobs for wages and essentially volunteered their time for union business until stipends were provided for officers. Employees had the right to file complaints over pay, seniority, and working conditions and have them grieved. Grievance committeemen were also given stipends for their service. Archived annual financial audits document the steadily rising cash balances and salaries of officers and grievance committeemen. In the early '60s salaries had become one of the items of union business, and a salary committee was formed to investigate appropriate compensation by comparing salaries among the various locals in District 4. Total compensation for all officers and grievance committeemen in Lockport at that time was approximately $6,500.

The minutes from the early union meetings in the 1940s are not among the records in the Penn State archives, which only contained letters of interactions between Lockport, Buffalo, and Pittsburgh. Early bargaining issues involved wages, hours of work, overtime, vacation pay, the Christmas bonus, the check-off system for dues collection by Simonds company officials, voluntary vs. mandatory union membership, seniority rights, and working conditions. These issues would repeatedly occupy union officials and management negotiations in future decades, but initially they were covered in a 22-point plan submitted to plant manager, Allen Potts.

Regarding wages for the rank and file, assistance was requested and received from union locals in Utica and Syracuse (District 3) to share their wage rates for similar work performed at Crucible and Allegheny Ludlum plants. The district men not only complied but shared their contracts. The basic wage rate at the time was 78 cents an hour. Based on feedback, Simonds' rates were adjusted and back pay was awarded in some departments. An annual one-week vacation after one year's employment was proposed. Potts proposed a five-year alternative for the one-week vacation, and he agreed to two weeks' vacation after fifteen years of employment. As the negotiations were taking place during the war, he also argued for adjusting the vacation period between June and October, or canceling it entirely and paying wages if vacations interfered with wartime production.

By May, contract negotiations had reached a deadlock, forcing District 4's Joseph Maloney to send an urgent telegram to the U.S Department of Labor's director of conciliation services. "Send a conciliator immediately as a very serious situation exists," it stated. Subsequent requests for conciliation would occur in the years after the war. A work stoppage occurred as summer approached. This caused considerable consternation and generated much correspondence. When Simonds found they had been left off a national preferred list of contractors, George Male sent a telegram to the director of the War Production Board imploring, "Do what you can to help," because it hampered hiring. Contract negotiations resumed, sticking points were resolved, and production stepped up; but 1943 closed out with the company's refusal to pay the annual Christmas bonus in December. Another appeal was made, and this time the dispute was resolved following a directive from the NLRB.

Union business in Lockport over the next few years involved construction of contract language for job qualifications and the evaluations of certain jobs by industrial engineer consultants. The union and management were at loggerheads regarding an outside consultant. Implementing new wage rates in terms of production versus maintenance, working out incentives for certain mill workers, conditions under which seniority could be lost and settling grievances through proper procedure, kept union officials occupied when they weren't arguing about selecting and then electing annual slates of nominees. As one might imagine, a lot of this activity did not go smoothly. Monthly meetings could be vitriolic, with much shouting and recrimination. Wages, union dues and eligibility for office were concerns and recurrent issues at those meetings.

Murray attempted to optimize steelworkers' wages from the very beginning of his union's victory in 1942, but they were frozen at 1941 levels through the war years. After the war, steelworkers clamored for higher wages, and steel companies wanted price hikes. The government attempted to have its oversight continue on permissible steel prices and hosted three-way conferences at the White House. The conference attendees reached an impasse that culminated in a strike in 1946, which initiated a series of wage and price hikes that spiraled into the next three decades. These gains did nothing for productivity but were seen in blue-collar households as a very good thing. Allen Potts was still the plant manager during the strike, and Edward Cook was president of the local union, while George Male continued to represent District 4. Whatever went on in Lockport, the 1946 strike was settled at the national level.

On the heels of that settlement, the Taft-Hartley Act of 1947 was passed as a major revision of the Wagner Act, after much resistance from labor leaders, congressional debate (John F. Kennedy and Richard M. Nixon were freshman senators on opposite sides of the issue), a veto from President Truman, overturned by the House and Senate. The act, which had many provisions, retained paragraph 7(a) of the Wagner Act allowing collective bargaining, but it restricted the activity and power of labor unions. These restrictions would come into play during the Eisenhower administration.

The result of the 1946 strike was an 18.5-cent-an-hour wage hike; steel companies won an average $5.00-per-ton price increase. The steelworkers' settlement was the basis of the United Auto Workers settlement that year, as steelworkers became the pacesetters for blue-collar wages. Additional wage increases occurred in 1947 (12 cents) and 1948 (13 cents), but the 1950s were the boom years for both wage earners and steelmaker profits. Steelworkers, whose families had endured low wages during the war compared to workers in other defense-related and auto industries, began to earn middle-class incomes and live a middle-class lifestyle, purchasing homes, automobiles, and saving to send their children to college. The 1950s would see the USW involved in several costly steel strikes for still better wages and benefits.

The first of these began in April, 1952, in the midst of the Korean War, and in the era of McCarthyism. The USW struck U.S. Steel and nine other standard carbon steel manufacturers for higher wages in what would become one of the most contentious and famous strikes in steel labor history. President Truman, responded, not by invoking the Taft-Hartley act, but by seizing control, i.e. nationalizing the industry to ensure the availability of combat steel. The steel companies sued, and the case worked its way quickly through the lower courts to the United States Supreme Court, which declared the President's action unconstitutional. Negotiations resumed between the union, companies and government officials leading to an end of the strike in July on terms that were similar to those proposed by the union four months earlier.

The union considered the strike a significant win, but it had serious economic losses for the country in terms of lost industrial output and unemployment—it had idled over fifty thousand steel and downstream workers in New York State alone, primarily in the Buffalo-Niagara region.

16
BOOM

In labor negotiations at the national and local level, patterns and substance could differ substantially. Some producers and local unions could break ranks or opt out, especially when issues in one segment of the industry were not pertinent in another—case in point—the specialty steel and the standard-carbon steel segments in 1952.

While steelworkers were on strike and three-way talks between Phillip Murray's USW, the big steel companies and government mediators were ongoing in the spring of '52, Simonds continued to operate, as a parallel and independent set of negotiations were conducted in Lockport by Simonds company executives and Local 2857 officials. No settlement had been reached locally when the big steel settlement was signed in July. That strike ended after fifty-three days when UAW President David McDonald (who had succeeded Philip Murray following a fatal heart attack), the big steel companies, and the government mediator—at the urging of President Truman, settled on a 21-cent wage hike, a $4.50-per-ton increase in the price of steel, and rather significantly, the establishment of a union shop policy, which mandated that new workers had to join the union.

In Lockport, the union and Simonds management were far apart on offers, but work continued as negotiations extended into the following year when the contract between union 2857 and Simonds was set to expire. In February the pact was extended as talks continued through the winter, spring and summer.

Ed Koleck was a highly-respected Simonds worker who headed up the union's negotiating committee. He had sharp labor negotiating instincts, so much so that Simonds later on hired him into their labor relations office. Eventually succeeding to the union presidency after Mullaney, he was also elected to terms as financial secretary and grievance committee man.

Koleck served as a World War II B-24 Army Air Force flight commander before coming to Simonds. He flew forty-seven missions, earning the European-African-Middle Eastern Campaign Medal with eight stars, the W W II Victory Medal, the Distinguished Flying Cross, and the Air Medal with Three Oak Clusters. After the war, he was a liaison officer with the Air Force Academy, and retired as a Lieutenant Colonel in the Air Force Reserve. The French government presented him with the French Legion of Honor shortly before his death in 2013. Characteristic of the men of "the Greatest Generation," he spoke little of his experiences and few knew of his exploits until his obituary was printed. The same could be said for other WWII veterans who worked at Simonds, such as Anio Nubelo, who participated in the landings in Sicily and Normandy with the U.S. Navy and received a number of service medals. After the war, Ed attended Cornell University for a time before he was hired at Simonds, where he began work in the labor gang.

Unable to reach an agreement, the 1953 strike began, on September 2nd, idling some 600 workers. (Meanwhile, all of the other steelworkers in New York State and around the country were back at work). The meeting atmosphere during strike negotiations was tense, in part due to a rupture in the ranks of the union steelworkers themselves. The situation was created by a small group of insurgents in a back-to-work movement who defied the majority and challenged the union president. This was a needless distraction for Ed Koleck's negotiat- ing committee, including the union officers, Mullaney, Rinaldo, Price, O'Donnell, Zimmer and Ramer. These men had several negotiating sessions with Simonds' Lockport management—Potts, Richards, and Ferris; however, day 45 ensued with no progress in sight. The talks eventually included District 4 representatives from Buffalo and Simonds executives from Fitchburg who met with U.S. Department of Labor mediator, Clarence LaMotte, at fractious meetings in the Lox Plaza and Park Hotels. Union and management had been far apart on the company offer of a 8-cent-an-hour increase plus a 5-cent "inequity" adjustment, and the union's demand for wages that had been negotiated for the big steel workers the previous year. Basically, the union had long aspired for wages that were at least similar to those being paid in comparable plants making specialty steel, such as Allegheny-Ludlum in Dunkirk, New York, and at Crucible in Syracuse, New York; but they never got them.

As the walkout continued, my father was able to maintain a decent income in construction on an Erie County section of the New York State Thruway. (In those days steelworkers did not receive strike pay, although unemployment benefits kicked in after seven weeks) Most of the other workers were able to find other jobs as well.

The strike for the Simonds workers ended after sixty days, just before Halloween, when Koleck announced that the members of the union had voted to accept the latest company offer—an 18-cent wage increase (including 9 cents to correct various wage inequities), the union shop clause, incentive pay for the 16 inch mill, and health and insurance benefits.

During Koleck's subsequent term as union president, copies of his correspondence with the District office in Buffalo and International office in Pittsburgh manifested his feisty nature. To David McDonald, President of the USW, he wrote protesting a $2.00 increase in dues to $5.00. (The initial dues were $1.00 in 1943.) A twelve-page attachment listed the signatures from the entire Simonds work force. He also complained that the annual convention, which was held in Los Angeles that year, was too distant and expensive for local delegates to attend. He advocated for future conventions in a location more convenient for the majority of the nation's steelworkers. To I.W. Abel, the secretary-treasurer and future president of the USW, he penned a lively handwritten letter complaining of "the run around" he was getting regarding a receipt for the dues sent to Pittsburgh. The check-off system of union dues collection by the employer had always been controversial, and record keeping was a concern. *"I'm getting tired of this run around and I want action right now. As I have said before and am saying now for the last time . . . Simonds Saw and Steel wants a return copy of R113 . . . and so does Local 2857. . . . I don't think it will hurt any of you to do what I have asked . . . Thank you!!!"* I assume Ed and Simonds got their receipts.

War-devastated economies in Europe and Japan were recovering in the 1950s, significantly increasing steel production with newer and better technology. The United States' steel industry was at its peak in 1957, accounting for a third of global steel output, but a global recession the following year led to a drop in consumer durable goods, which impacted the steel industry with a plunge in steel orders. Regionally, 20 percent of the labor force at Simonds was laid off and work in various departments curtailed; some were on 32-hour weekly schedules. The industry nevertheless responded by raising prices. Interestingly, despite the economic downturn, 1959 was a big year at Simonds in Lockport. Major additions included a new press shop that year, and bar mill finishing, and a metallurgy lab expansion followed in 1960. There were also cold rolling, annealing, and heat treatment furnace additions.

In the midst of these local expansive projects, a nationwide 116-day strike in July, 1959, followed failed negotiations between David McDonald's USW and a joint effort by Bethlehem and other big steel companies. The executives of the big steel companies visited the White House to complain about their rising costs. They were willing to risk a long-term strike and asked the president not to intervene. Five hundred thousand workers were idled, and as the strike dragged on, the Eisenhower administration felt they had to get involved. The Cold War was still on, and the president's advisors had earlier considered "steel to be the most strategic of all strategic industries." In October, the president invoked the Taft-Hartley Act's cooling-off provision, sending disgruntled steelworkers back to work.

The steel strike in his last year in office was just one of several crises and incidents in Eisenhower's second term (Suez, Hungary, Little Rock, Sputnik, Gary Powers' U2 flight, Khrushchev and Castro). Consumer prices had not changed, but the steelworkers' union, eying earlier steelmaker profits, wanted to increase the wages of its workers. For their part, management wanted changes in the work rules. At issue was paragraph 2(b) of the master contract, the past practices clause, which steel companies felt impeded innovation and productivity. It regulated the number of workers on a particular job—the "featherbedding" issue—and basically stated that the number on a job like the sheet or bar mills could not be reduced unless the company introduced new technology. The rules also defined how the workers in one department were or were not able to work in another department. This management tactic aimed at job reform had been successfully beaten back by the union time and again. The strike was settled when President Eisenhower was out of the country on a continuation of his global goodwill tour, and Vice President Nixon and the Secretary of Labor convinced the steelmakers and the union to relent some of their demands. The settlement did not involve a change in work rules, but it led to an increase in hourly wages, an automatic cost-of-living wage adjustment, and improved pension and health benefits. Many observers of the industry, journalists and historians alike, believed the strike of 1959 was devastating in the long run by opening the door to imports from overseas producers who had improved their productivity. But price and wage issues were only one of the factors that contributed to loss of the competitive advantage that the American steel industry had enjoyed.

At the time, with an average wage of $3.10 an hour, Simonds provided one of the best blue-collar jobs in town. I had begun my summer job at Simonds in June 1959, the month before the big nationwide strike. I was totally unaware of the strike. In fact, Simonds hired workers like me, Gordon Martin, and others during the strike, which again pointed out that USW activity involving one segment of the industry did not necessarily impact the other—the specialty and standard carbon steel mills were different animals. But I was still a teenager then, happy to have a good-paying summer job, and I wasn't thinking very deeply about labor issues.

Labor pacts between unions and management were negotiated on a regular basis about every three years in the auto and steel industries. I witnessed the hard-driving unionism of Detroit when I lived in Ann Arbor during the 1960s. The UAW would target one of the big three automakers, and the other two would eventually join in the settlement. The forerunner was the USW. Between 1946 and 1959 there had been nearly a dozen contract negotiations. Sometimes the bargaining, at least at the national level, seemed reasonable; other times it resembled ransom demands after hostage taking. The union walked out half the time.

■　■　■

The 1960s saw the United States steel industry's dominant grip begin to loosen, although some steel historians believed the country had begun to lose ground long before. In 1962, the USW and steel companies negotiated a modest increase in wages. On its heels the steel executives created a major political storm by raising prices, upsetting not only the union who felt they were cheated out of a larger wage increase, but Kennedy administration officials who were taken by surprise. Steel prices were a significant factor in the American economy, and Kennedy feared the price increase would fuel inflation. Political pressure was placed on steel executives to roll back their prices, which they did, but the result was that "everyone looked bad," Stoddard wrote. As the decade progressed and the U.S. stood pat in its pre-WWII methodology, imported steel from Japanese and European companies that had made major technological advances, increased six-fold.

The United States had emerged from the war as the world's largest steel producer, but the Europeans refused to stay down. They continued to innovate as they rebuilt their steel industry. Canada introduced the basic oxygen furnace (BOF) to the U.S. in 1954. It was invented in Austria in 1952. Oxygen blown into molten iron could complete a heat more quickly and efficiently than the open-hearth process. The BOF began out-producing the open-hearth by the 1960s.

In 1962, the integrated steel mill began evolving into the automated factory of today, first with the continuous-casting process introduced by a Roanoke company, whereby molten steel from the furnace was tapped into a ladle and then directed into a casting machine and rolled to a finished product without the various and labor-intensive manual operations. No one gave much thought to assembly line automation at Simonds. Specialty steel made in small batches was more of a customized operation, and much of it remains so today. However, in modern integrated mills, the making of molten steel in one of a myriad of customized chemistries, is directly linked to casting and rolling that is untouched by human hands. Operators sit high above the factory floor in glassed-in booths at computers, manipulating state-of-the-art automated equipment.

During the '60s, as as the first experiments with automation were getting underway elsewhere, Simonds continued to do very well with its electric furnace and manual methods—primarily because of the many defense and other industry contracts it had for its specialty alloys. The plant operated twenty-four hours a day on three shifts, six days a week. Despite a seemingly good bottom line, in 1965 the Simonds family decided to sell their shares to the Wallace-Murray Corporation, ending 133 years of family ownership. The Simonds Division became a part of a newly formed New York City-based industrial conglomerate. In the years that followed, the company began to experience competition from lower-priced foreign specialty steelmakers. In one of his earlier forages through the plant, Michael had come across a literal sign of those times when steel executives were lobbying for and getting protectionist legislation: *Protect Your Job, Buy American*, it read. While business was booming, my father and most of the other employees continued to work steadily.

In Lockport, Local 2857 maintained "reasonably satisfactory" labor relations with the Wallace Murray Corporation as they had with the Simonds family, despite the fact that Simonds executives had been strongly opposed to organizing their Fitchburg plant. In 1963, G.K Simonds, the company president, wrote a letter to his Fitchburg employees urging them not to sign a union card. "We are convinced," he wrote, "the employees in this plant did not need a union in the past and do not need a union now and that the overwhelming majority do not want a union."

Mathew Armstrong succeeded District 4's longtime Director, Joseph Maloney, in 1965, the transitional year from Simonds to Wallace Murray. Correspondence between Armstrong and District 1 Director, Roy Stevens, representing Fitchburg, urged Local 2857 to support Fitchburg organizational efforts by providing copies of their contract booklets and letters of encouragement citing union gains in working conditions. Invitations were sent to Simonds' local representatives to attend organizational meetings in Fitchburg in the hope of awakening all parties to the "realities of coexistence." The process took several years, finally coming to fruition in 1971, nearly thirty years after the Simonds plant in Lockport was organized. The Lockport local made a similar effort to support a successful campaign to unionize a Simonds Saw and Steel operation in Portland, Oregon.

Meanwhile, Local 2857's monthly meetings of the rank and file, and the executive board's deliberations were often spirited during the tumultuous '60s. The local's president during much of that decade was John Tice, while the manager of Simonds Labor Relations was R. K. Bishop. David Mackey had succeeded Allen Potts as the new plant manager in 1960.

Many had expected Don Richards, the plant superintendent, to replace Potts. But the Simonds executives in Fitchburg decided that Richards was too valuable in his supervisory position and would be difficult to replace. He knew the workers and managed them well, usually making sound, wise decisions that the union respected. Allen Potts was a capable manager with a formal presence in the office, but he rarely entered the plant or spoke to the workers. Mackey, who stood a head above most of the steelworkers who surrounded him in photos taken at company functions, was an affable and capable administrator who was well liked by personnel throughout the plant. A walk-around manager like Richards, he also seemed to know the steelworkers by name and was respected by union officials.

Wage policy was only one of the issues confronting the executive board. Union dues, initiation fees, and exoneration payments, controversial grievance committee outcomes, NLRB inquiries into denial of grievance hearings, candidate eligibility for office, disputed elections, civil rights complaints, and labor disputes culminating in work stoppages were just a few of the other issues that required a great deal of time and effort. Meetings held at the Lox Plaza Hotel were remembered as boisterous and chaotic affairs.

One meeting that John Coleman told me about involved two workers protesting denial of a grievance committee appointment. The meeting was so disruptive that the union was forced to bring a police officer to keep order. The weekend afternoon meeting was well attended by union members who, now dressed now in casual clothes, slacks and sport shirts, craned forward in their seats, startled by an angry harangue between the two disruptive workers who were slandering the officers at the head table with abusive language. The incident resulted in some of the officers resigning before reconsidering their decision following a vote of confidence by the membership. The two workers who had slandered the officers subsequently received formal letters sanctioned by the district office and the national union warning of possible dismissal should they fail to manifest civil behavior in the future.

Another strike in 1968 was settled when Mackey and Tice's local agreed on a significant increase in hourly wages phased in over a period of three years. Mackey passed away suddenly of a heart attack in 1969, and was succeeded by Chuck Emery, who had been in charge of sales. The Potts-to-Mackey-to-Emery transitions went smoothly.

The peak year for employment at Simonds was 1969, when nearly 1,000 workers and 100 salaried staff were on site. The plant was booming. Orders for sheets, plates, and bars of tool and stainless alloys for aerospace, appliance, electronic, residential, and commercial construction were at record levels. The company couldn't hire enough workers to meet demand, despite advertising for workers well beyond the Northeast, even into greater Appalachia. One of the men remembered a father and son moving up from that region to fill Simonds openings. They were two of only a few recruited from outside Western New York.

When Simonds was at its peak in output, and unemployment was close to zero, those agreements over a new contract became bogged down as the old contract was due to expire. The new union president, Herbert

Starr, petitioned USW president, I.W. Abel, for permission to strike. Abel consented, but the strike was averted when Starr and Emery agreed to an indefinite extension of the labor pact negotiated three years earlier. Averting strikes rather than dealing with root problems was a commonplace negotiation strategy throughout the steel industry at that time.

Not long after the new contract went into effect, the famous "sick out" occurred at Simonds. The incident began with the filing of a written grievance when management decided to replace two union workers' jobs with salaried employees during a slow point in the business cycle. This action resulted in the entire work force calling in sick, shutting down the plant as the grievance process was worked through. Whether it was the rule itself or the personalities of the replacements is conjecture, but in the end, the union workers returned to their jobs. The whole incident was "plain foolish," Dave DeLang recalled. It was foolish because the rules in the steel industry had long established that management personnel could not perform jobs that had been assigned to the union. It was a big decision for a union worker to leave for a salaried position within the company, because once he made the move, he could not return to a union job with that company.

Nineteen seventy-one was a good year for most American steel companies and workers; but as the decade progressed, foreign competitive advantages grew, and the winds that buffeted all steel companies were accelerated by the added stress of nearsighted actions by union and management as new forms of domestic competition became manifest. The predators had been out there for some time studying their prey, but they had kept their distance, waiting for their opportunity.

The Experimental Negotiation Agreement of 1974 gave the union cost-of-living increases in exchange for pledging not to strike; however, actions like that came too late. It had become cheaper to import steel, which was what both steel companies and downstream users began to do.

17
BUST

The drama of steel's decline began to play out in the cities of the large high-carbon steel producers: Pittsburg, Baltimore, Buffalo, Cleveland, Chicago, and dozens of other towns of the so-called "little steel" group: Republic, National, Youngstown, and Inland. Initially, the specialty steel alloy companies like Simonds Saw and Steel in Lockport did not sense the ominous changes that were about to take place for them as well. By the late 1970s, a third of Simonds' sales volume was supplied by foreign sources.

The root causes of the decline were many and complex, but among them were that the powers-that-be in the steel industry and labor were unwilling to change their thinking or their ways; and government actions during the Reagan and Volker years did not help. As Mark Reutter argued in *Making Steel,* "They [industry] spurned product research. They choked off innovation. They relied on inflating the price of steel to keep up quarterly dividends, and they let the situation deteriorate to the point where their mills were out of date. In so doing, they created the opportunities for their rivals—not only for the Japanese and other foreign steelmakers, but for domestic "mini mills" and makers of steel substitutes, such as aluminum and plastics." Pre-stressed concrete was another product that substituted for the steel girders and beams that formed the framework of many new commercial buildings. John Strohmeyer, author of *Crisis in Bethlehem: Big Steel's Struggle to Survive*, reached a similar conclusion: "a failure of vision, a lack of interest in new methods and efficiencies, and an indifference to customers," he wrote.

Labor costs played a big role as well in what had become a highly competitive marketplace where high rates of return were expected. Union demands for higher wages, more benefits, and working conditions that favored less-than-optimal efficiencies, took their toll. Attitudes about work began to change in the 1960s, more than one Simonds steelworker told me. The new hires didn't want to work as hard as previous generations. This attitude was not limited to Simonds. Deborah Rudacille writes in her memoir that similar attitudes existed among workers at Bethlehem, who passed their attitudes on to new hires. In her interviews with active and retired company men, union executives and workers, she learned that the company implored the union to urge their workers to be more productive; that there was too much featherbedding. Work rules had evolved to the point that jobs had more people assigned than were needed. It was too easy to create jobs, and foremen were stymied in their attempt to make the crews more efficient. "There were workers that would file a grievance at the drop of a hat and the union would support them." Grievance committees spent a lot of time protecting less-than-optimal workers. I heard similar frustrations from retired Simonds foremen. When a worker attempted to pitch in and help out a fellow worker, the union rep would remind him of the rules: "Do your job and nobody else's."

Labor journalist, John Hoerr, blamed both labor and management in his book, *And the Wolf Finally Came*. Hoerr cited costly confrontations and miscalculations between labor and management. The crisis in the steel industry was preceded, he wrote, by "decades of adversarial relations on the shop floor." In his account, each side pursued their own goals with selfish disregard for the other. He took management to task for failing to recognize the threat of foreign competition, upgrading plants, introducing newer technologies, and for alienating workers. He found fault with the union for protecting work to the extent that efficiency and productivity declined at a time when they were increasing in the Japanese and European steel industries.

In addition to the foreign competition, the big steelmakers were faced with a new kind of domestic competition, "the mini mill." It began with a North Carolina company called Nucor. The company produced steel bars, bolts, rod, and wire items at lower cost than the big structural steelmakers by using electric arc furnaces to melt scrap metal. The company had an interesting history of transitions from automobile manufacturing to nuclear services and finally to steel; a sequence that involved bankruptcy, liquidations, reorganizations, acquisitions, and relocations in four states over more than a century. Tracing linearly to the manufacturer of the Oldsmobile, Ransom Olds in 1897, Nucor achieved its initial success in the 1960s under Kenneth Iverson, a trained metallurgist, and Samuel Siegel, the chief financial officer. Iverson began making steel with the electric furnace, later adding the vacuum furnaces, a continuous caster for converting molten steel to solid form, and re-heating furnaces, sheet, bar, and plate rolling mills. Nucor hired non-union workers, did not offer defined benefit pension plans, and gave workers less-generous health plan options. The company paid good wages, but avoided the expensive legacy costs built into pension and benefit plans, instead relying on fewer highly motivated workers, incentives, bonuses, and pride in work. Management structure had fewer layers, translating to better strategies and faster decision-making. Beginning at one-fifth the size of Bethlehem in the mid-eighties, Nucor marketed steel at 15 percent below Bethlehem's prices, captured much of Bethlehem's market, and surpassed them in sales by the year 2000. As the new millennium continued, executive leadership moved from Iverson to Dan DiMicco as they made several steel and raw material company acquisitions and introduced the "micro mill." In 2013, John Ferriola became the CEO. Additional steel companies and rebar facilities were acquired. They added to their processes marble-sized pellets of direct-reduced iron to the mix of steel scrap to raise the purity of the melts. The relatively low cost of setting up their mills allowed them to compete not only with U.S. Steel and Bethlehem for high-carbon steel, but with the alloy steel producers as well. Soon they began making inroads in contractual arrangements. Today, Nucor is America's largest steel producer, the largest mini mill, and the largest recycler of scrap metal, largely from junked automobiles.

Nucor, initially a high-carbon steelmaker, was not a Simonds competitor; but in the 1970s, Wallace Murray management attempted to respond to the competition it did have with Nucor-like efficiency in order to cut cost. Technical adaptation and removal of positions, for example in the 10-inch bar mill, where the assistant heater and other positions were eliminated, reduced the crew size from twelve to eight workers. But Nucor and Wallace Murray cultures were not the same; the sorts of changes they made at Simonds had minimal effect on the bottom line. When the local controller suggested more significant budgetary changes to the Wallace Murray corporate controller in New York, that changes in management were needed, the corporate controller said, "We don't fire management, we sell the division." Not long afterwards, that's what happened.

I spent a month in Lockport in the summer of 1973 transitioning from the medical corps at Bethesda Naval Hospital to the start of my practice in Arizona. I was able to spend time with my father because he was not working as regularly as he had been. The sheet mill had gone full-tilt with lots of overtime in the '50s and '60s, with three shifts roll-ing out tons of sheet steel. Much of it was shipped to Fitchburg, where it was finished into saw blades. One of the bigger orders for hacksaw blades was shipped from Fitchburg to Harrison Radiator in Lockport; but in the early '70s Harrison canceled their Simonds contract in favor of cheaper Japanese blades. Other contracts were lost as well. The overtime disappeared, and three shifts were reduced to two, then one, and even work on that shift became erratic.

In the old days, extended time with my father other than Sundays was unusual. Even then, Sunday afternoon, after Mass and our traditional pasta dinner, was naptime. He usually worked one of his two weeks off in the summer during Simonds' annual shutdown doing maintenance and refurbishing work at the plant. During the second week he would do a remodeling project or paint the house. We frequently drove to Canada to visit my father's four sisters and their families, who had immigrated there in the early '50s. One of them lived in the Brightside neighbor-hood of Hamilton, Ontario, near the big Steelco Company steel mill, a neighborhood that was eventually gobbled up by the plant's expansion. My father and his brothers-in-law had been able to discuss steelworking at Simonds, Steelco, and Dofasco, Hamilton's other steel mill; but now, in the late '70s, he had little to say on that subject. He seemed bored. The job at Simonds had been his trademark, and he had relatively few outside interests. He had planned to work to the customary retirement age of sixty-five, but each year the periods between workdays or weeks lengthened, and finally in 1977, at the age of sixty-two, he retired. I was actually happy for him. The life of a steelworker could end suddenly from an accident or slowly from work-related cardiac and other chronic diseases. Decades of labor in the mill had taken its toll on the previous two men who had his job—they had heart attacks. I didn't want that to happen to him.

My father adjusted. He was able to collect his pension and occupy his time visiting with friends and relatives, putter around the house and yard, and make occasional trips to Canada and the race track. Meanwhile, my mother continued to work at Norton Industries, the plastic factory in town, for a few more years to improve her pension and Social Security positions. Ironically, plastic was one of the materials that made inroads in building products, automobiles, and household goods, territories that had once been the domain of steel. Simonds didn't make automobile strip or frame steel, but they did make steel for auto parts and dashboards. Research and development in plastics had paid big dividends by developing thermoplastics and other types that could be used as cheaper substitutes for steel in certain applications. Plastic was lightweight and could be easily molded into a variety of products. My mother worked in Norton's mold shop, cranking out things that were once made of metal. Pre-stressed concrete and aluminum were the other products that the steelmakers didn't realize would replace a significant segment of their industry. Pre-stressed concrete began to replace steel beams and trusses for buildings and short-span bridges. Steel, after the development of a specially coated lining, accounted for nearly 100 percent of beer, soft drink, fruit, and vegetable containers—but lost nearly all of the market to aluminum. None of that was Simonds' business, but foreign and domestic competition did affect them.

As difficult as it was for senior workers like my father, the downturn and layoffs at the mill were devastating for the younger men who lacked seniority. Friends and relatives who had graduated high school, served in the military and had come home to a good job that they thought would get them to retirement with a good pension and benefits, had a hard time adjusting to the new realities. The average age of the men at the time the plant closed was around fifty. It was difficult for men in their fifties to find other jobs. "You looked for something good, but nobody wants you," one of them told me. A precious few were able to get jobs in other steel mills, but they had to move to other cities and states, because all of the Buffalo area mills were suffering the same fate. However, those opportunities dried up as the steel industry decline became more widespread. Salaried workers with college degrees, and steelworkers who had started taking college courses at night while working days at Simonds and who had obtained degrees, were able to find work in other metal industries. A few became entrepreneurs, establishing businesses based on the expertise that they acquired at Simonds. Others had technical skills or aptitudes and were able to retrain at vocational institutes for welding and other jobs, but many couldn't. When unemployment ran out, steelworkers became bartenders, used-car salesmen, real estate agents, city and county workers, correction officers, and warehouse managers, while others spent long periods scratching for whatever minimum-wage jobs they could find in the emerging service sector. The impacts were felt not only on steelworker families and neighborhoods, but in other blue- and white-collar jobs as well. Workers at the Harrison Radiator Division of General Motors, the city's largest employer, began to experience layoffs due to the downturn of the automobile industry. All of this added to the despair of what was happening at the historic downtown core of the city.

In the first decade of the nineteenth century, the Lockport area was a forested wilderness. The first settlers arrived in 1815–1817: The five Comstock brothers were among the mostly Quaker group of fifteen men who owned most of the Lockport area property. Only three families had carved out a homestead by June, 1818; however, by the end of the year, 350 people lived there. With the Erie Canal dig underway, several hundred more families had moved in; and by the time canal construction was completed, 3,000 families were living there in addition to 2,000 Irish immigrant men. Lockport grew to a small city of 25,000 in the mid-1900s and flourished as a commercial and industrial center in Niagara County.

Main Street was the heart and soul of the city in the 1950s when I grew up there and worked during my high school years at the downtown Rexall corner drug store. Lockport was thriving, full of life, the shops and businesses nearly all locally owned except for some of the banks. There were more than seventy bars and restaurants, many in the core, clamoring with people. Parking was easy. Everything bustled, it seemed, and there was nearly full employment. When you wanted to shop or meet someone you went downtown, not out of town.

The Erie Canal divided the community into two regions as I saw it. The more fashionable parts with the nicer homes and neighborhoods were east and south; the more modest homes and ethnic neighborhoods were west and north of the canal. Simonds and most of the other manufacturers, where the American dream took hold in Lockport, were located west of the canal on its more southern and northern ends. However, it didn't matter on which side of the canal you lived or where you worked, it was a great town to start a promising job and to raise a family.

Then in the mid-1960s, as the omens of industrial decline were about to become manifest, the city began a precipitous decline that resulted from a series of poor decisions connected to a federal Urban Renewal program. The distinctive architectural character of the Main Street shops was bulldozed to make way for a new urban plan that was poorly executed over time in a piecemeal fashion. Many businesses that had helped build the town, such as the Williams Brothers Department Store, left or closed. By the early 1970s, the downtown demolition scene along Main Street resembled a World War II bomb-damaged city. Stagnation set in as conflicts between city officials and developers were litigated. At the end of the decade, as the global economy was beginning to have its effect on the local economic landscape, it was apparent that the Main Street project was failing, while over on Ohio Street, Simonds was beginning its downhill course. Local government, company, and union action that might have created long-term success for everyone, even in the global marketplace, failed to develop. Policies to preserve the urban core, to attract new business to keep its citizens downtown working and shopping and young people from moving away, never materialized. Maintaining civic integrity in the face of a failing economy was and remains a challenge. Some towns in America with older but intact commercial districts rose to it; others, stymied by complex political and economic interests and issues, wouldn't, or couldn't do it.

A year after my father retired, in 1978, the Simonds division of Wallace-Murray was sold to Guterl Specialty Steel, a Pittsburgh concern. In the years following their purchase of Simonds, Wallace Murray performed well. A 1968 memorandum stated that sales and net profit had tripled as a result of "internal growth and acquisitions, particularly Simonds Saw and Steel." But a decade later, Wallace Murray saw the writing on the wall. They did not continue to get the return on investment they had expected, and believed they could buy more-profitable operations. Had the Guterl owners and investors done their homework, they might have realized it as well.

Guterl continued producing various specialty steels, but the key to its survival lay in maintaining its defense contracts, according to its president and CEO, Douglas Pinner. But Guterl, like other steel producers, was not immune to the root causes of the steel industry's decline in general. The Lockport plant was largely antiquated, although some departments had up-to-date and functional equipment. In many of the plants of the major steel producers, physical constraints, lack of capital, and management mentality prevented efforts at modernization and cost savings that would have accrued, had such technologies as the basic oxygen furnace and continuous casting been installed. But those innovations cut across the major phases of steelmaking and several mill departments where supervisors were reluctant to relinquish control. Those constraints, which impacted the big integrated mills such as Bethlehem, were not at play at Guterl, but new technology was badly needed. At the same time, there were competing management and steelworker demands. The steel strikes of the 1950s and1960s had established work practices that increased costs and helped to facilitate foreign competition.

In 1980, two significant events followed on the heels of one another. Union local 2857, whose president was Jim Geier, along with officers, Carlo Snell, Terry Forgie, Alfonse Rosati, Bill Winters, and Louie Koel, sat down to negotiate a new three-year contract with the Guterl team. It would be the final contract in Simonds-Wallace Murray-Guterl history. Interestingly, Herbert Starr, who had been the local president years earlier, advised Geier's group as the USW's representative from District 4, while previous 1950s-era local president, Edward Koleck, took part on the Guterl management's team as its labor negotiator.

There was the usual posturing early on, each side presenting its positions on the issues, and the union implying a strike might occur if its demands were not met. Contract language and process at the district and local level had replaced industry-wide bargaining, which had both positive and negative aspects. Wages, benefits, work rules, and overtime were the usual items that were worked through. Established rules in the integrated mills kept men in certain jobs and departments. At Simonds, from the management perspective, getting the right personnel in the right job, and worker progression, especially on overtime and in the case of illness and accidents, was what mattered most. Department level seniority was king—it's what mattered most when jobs were posted— plant seniority was secondary. When cutbacks occurred, the low man on the department totem pole joined the labor gang if he was not laid off entirely. In the labor pool, plant seniority rose to prime importance among workers vying for reassignment to various departments. When overtime on a job came with big financial incentives, the competition and emotions could be intense. Bumping rights could be a thorny issue. The union tended to support the seniority system, but the crew as well as management wanted workers who could also do the job well. Everything worked well when the senior man was also competent at a particular task, but trouble arose when a worker's performance was less than optimal. Then everyone else lost money and it forced job reassignments and workarounds. A worker could bid a job for which he had little or no experience. A specified period of job training was a contractual right; however, problems arose when a junior worker resented having to train a more senior man. Getting the language right in contracts made the difference between an easy progression and one that was more difficult. Both sides got down to the language nuts-and-bolts and reached agreement on a new contract in the final weeks of the 1980 negotiations.

The new labor agreement, which came soon after Guterl had received the news that it had received the final notice of a guaranteed 15-million-dollar government loan to modernize the plant, called for salary increases of 25, 20, and 15 cents over the three years of the contract, cost-of-living and insurance adjustments, and satisfactory overtime.

Everyone was happy, or so it seemed; but in hindsight, maybe the union should have seen what was coming and looked to better protect their pension and benefits. The reality, however, was that no one saw it coming; neither union nor salaried workers had a crystal ball.

The same was true at the other Western New York steel mills and foundries where about 20 percent of the region's workforce was employed. The Bethlehem Works at Lackawanna, which employed 10,000 workers, less than half its peak workforce, and the smaller outfits—Republic Steel, Shenango Ingot Molds, Hanna Furnace, Amherst Foundry, and Atlas Steel—were all in trouble. Some of the workers and union men sensed something was wrong, but only a few anticipated the plant closings that would occur in the near future.

In Lockport, the Guterl owners and local managers were also all smiles after the loan guarantee and new labor agreement. Company president, Douglas Pinner, had high hopes to upgrade operations and to double his work force, which had shrunk to 500 workers. The loan was used to purchase and install state-of-the-art vacuum induction melting (VIM) and electro-slag re-melting (ESR) furnaces in the existing melt shop space, replacing the electric arc furnaces and the argon-oxygen-decarburization units. Compared to the performance of the EAF and AOD, the VIM allowed for much more precise melting of steel, and the ESR process further purified the steel to the ultraclean requirements of aerospace industries. Pinner had also hoped to find increased efficiency in a Guterl innovation he referred to as the "bottom pour," in which molten steel was poured through a cylinder into the bottom of a series of five or six molds that filled from the bottom up. The method avoided side splashing and mold erosion while increasing the yield and improving the quality of the steel.

But then, in 1982, the bottom dropped out just as the new VIM and ESR operation got under way in late June. The recession, the loss of contracts and market share led to more layoffs and culminated in the filing of bankruptcy on August 6th, two months after Bethlehem announced it would close its Lackawanna steel works at the end of the year. The closure of the biggest facility of the country's second largest steel producer, the integrated mill that once employed 25,000 workers and was a sure sign that the American steel industry was dying. That year, the New York Times reported that the American Steel Industry was operating at half its capacity.

18
AFTERMATH

"**I** knew they were going bankrupt," Lou Valery said when I asked him why he decided to leave his job supervising the sheet mill department to take a chance on a startup steel business, Niagara Specialty Metals.

"What happened?"

"The union ruined it, to be honest with you," he responded. "They [the company] gave them [the union] everything they wanted." It may have been true that steel companies had over-given, but many steel historians also believed the union had overreached. Lou had started out as a worker at the plant in the late '50s, but by the '70s, he had become a foreman and often felt stymied by union work rules. "If they didn't want to work they could just sit there," he said.

Other salaried employees I interviewed would have agreed with him, while not absolving management of their failures. "The union got nearly all their demands," one of them told me, "but management paid no attention to the budget."

In Lou's opinion, the union took bargaining from the goal of collective representation in the 1930s with the aim of achieving fair wages, benefits, and working conditions, to one where a good deal of time was spent protecting workers who didn't want to work hard. The grievance clause in union contracts allowed workers to file complaints, which could be "grieved" through a five-step process. This worker protection had come a long way since its inception with the establishment of the USW in 1942. Even after the protection was in place, old-time workers were afraid to file a complaint for fear of losing their job. That was no longer the case.

At Simonds, the worker complaining to his supervisor or foreman could usually quickly resolve minor issues such as a particular job assignment. If that failed, he could get his union steward or president involved. The issue might be solved after the grievant took the problem to the labor relations manager, who had a few days to consult with management. But if that failed, major issues were written up and presented at scheduled meetings. At these meetings the union's grievance committee members would have a sit-down with company labor relations men. Depending on the grievance, the problem was resolved quickly at the first meeting, or, if both sides dug their feet in, negotiations would drag on for a while. The final step, if the issue could not be settled after good-faith sessions, was arbitration utilizing an outside mediator. This was a rare event; if it ever occurred at Simonds, no one could recall.

Lou said, "If they (any mill crew) had a guy on the grievance committee they could refuse to work over some lame excuse. You could go through the steps and write them up, but they would laugh at you, and they [management] wouldn't do anything. And there were more men than were needed. At Simonds [Guterl] it took 23 guys to roll a sheet of steel. At my place I could do the job with six." But Lou didn't have to deal with a grievance committee. He ran a non-union shop.

His business partner, J. Barry Hemphill, was determined not to have a union shop. Hemphill had arrived at Guterl in 1979 after a career at the Duquesne works of U.S. Steel in Pittsburgh, and had witnessed the effect of unions on productivity in both Pittsburgh and Lockport. With a degree in chemistry, he had moved up the ranks from a management trainee to general foreman of the melting operation at Pittsburgh. At Guterl, he became the supervisor of rolling operations. When he got his opportunity at Niagara, he fashioned a business structure modeled on Nucor and initiated a profit-sharing plan wherein the employees received 25 percent of the pretax profit. There were other incentives as well, which allowed the company to use fewer personnel to accomplish the same Guterl output. "You'd be surprised to see what a group of highly motivated men can accomplish," he told me.

Better worker and management relationships at Guterl might have helped stave off its precipitous decline; but, in truth, the animosity between management and the steelworkers' union, their wage and benefit demands, and work rules were but a few of several factors that had begun to buffet the steel industry across the country.

Steelmakers long complained about steel imports and lobbied the government for relief from foreign competition in the form of tariffs and quotas. Eventually, Voluntary Restraint Agreements (VRAs) were enacted during the Reagan administration. These helped some manufacturers for a while, but not enough for what was ailing many of them—lack of access to new markets and contracts, failure to innovate and match overseas and domestic mini mill technology, and inefficient working practices. The recession, created during Paul Volker's term as chair of the Federal Reserve Board, and also during the Reagan administration, dropped the demand for steel in general for both structural and specialty producers. Locally, management and labor issues became magnified, market share was lost, and red ink began to flood Guterl's financial sheets.

The Guterl owners, Bill Guterl and his associate, Mark Saxman, had already made their money in the integrated mills of U.S. Steel and Latrobe Steel. Although not living the lavish lifestyle of Charles Schwab with his grand Manhattan mansion and Loreto, Pennsylvania estate, his expensive collection of European art and his pursuit of women and gambling, the Guterl owners nevertheless enjoyed extravagance, and they had their own distinctive ownership style.

Jim Calos told me that when he reported to work early one morning, he encountered an attractive new woman in the office whom he had not seen before. "I'm a new employee; this is my first day," she said. The owners had hired a cocktail waitress for the business office while they were having dinner in a Buffalo suburban restaurant the night before. The spontaneous hire reflected many of the worker and salaried personnel's opinions of the executives: "They lacked discipline and didn't have a clue about how to run a steel mill." Under the Wallace Murray banner, management and union workers were a relatively happy family. They didn't always agree, but there was cooperation, and union workers respected the management of Emery and Richards. But "it all went away," one of the former salaried men told me. Emery bailed out soon after Guterl took over. He recognized that the new arrangements would not work. He returned to his previous company, Washington Steel in Pennsylvania. Jim Purviance became the plant manager, technically vice president of operations, during the Guterl years. More than one former worker indicated that shortly after the euphoria of the new contract and government-guaranteed loan wore off, things began to change. Management became more distant, and in hindsight, there was a suspicion that Guterl was shopping the plant to potential buyers.

"It was a zoo!" more than one salaried man told me. The bottom line of the bankruptcy and plant closure was, simply, "poor management," the reason cited most often by both retired union and salaried workers. As they said the words, they shook their heads, remembering the sadness of having to leave a place that had been such an important part of their lives, a place where they had so much enjoyed working. The Allegheny Ludlum president said as much when he assumed operations at the plant in 1984 and spoke to the local press. "This business failed because of the way it was run. There were many factors—poor management, low capital, and bad labor practice." Thirty years later, no one I spoke to disagrees with that assessment.

When the steel business was booming, steel executives could play fast and easy, spend lavishly, and do wild and crazy things; but not in the 1980s, when tight management and adherence to budgets were keys to success. The road Guterl took to bankruptcy was similar to the one traveled by other Pittsburgh-based big steel companies: a road paved by relative indifference to budgets, lack of foresight, and failure to adopt new methodology until it was too late. In Guterl's denouement, the gremlins were sensed, but the attempt to vanquish them with the VIM came too late.

The other factors that the Allegheny president mentioned: casual oversight and slipping discipline on the factory floor took their toll. In those last years, the result was poor-quality work and defective product, often knowingly shipped before full quality control was completed. Timing was another unfortunate factor. A recession had taken hold of the economy and brought with it decreasing demands for Guterl products in a year when foreign steel accounted for 22 percent of the market. The Simonds Company, which at one time employed nearly 1,000 workers in the Wallace Murray years of the early '70s, was down to 125 employees in the summer of 1982—from a high of 600 workers when Guterl took over operations in 1978. In the two years after the 1980 contract was signed, business, which had always been somewhat cyclical, was on a downhill course, and as midsummer approached it was in free fall. It became obvious to the owners and investors then that the solution to their financial crisis could no longer be postponed. On Thursday, August 5th, 1982, when the workers on the 7–3 shift clocked out, they didn't know if they would clock in again.

Earlier that day they stood dumbfounded before bulletin boards in the plant and read a tersely written statement: *On Wednesday, Aug. 4, 1982, Guterl Steel Corp filed in Pittsburgh for protection under Chapter 11 of the federal bankruptcy code. We intend to remain operating for the short-term and in the future. We will keep you informed as to specific plans and developments.* It was signed by John H. Hofstetter Jr., Vice President for Industrial Relations. On Thursday, August 5th, a pall hung over the security building's exit gate as the workers made their way to their cars.

"It's real depressing," one of the workers, a crane operator, mumbled to the *Union-Sun & Journal* newspaper reporter who, standing outside the exit on Ohio Street, recorded their comments.

"We figured this would happen sooner or later, all the other steel plants are going under. But it happened a lot earlier than we thought it would."

"No one knows what's going on yet," a mill hand barked.

"We'll know soon enough," another responded.

"It's just the way this company has been run these past few years," an angry worker muttered.

"No, it's just a steady decline in sales. It's the economy," said another.

Most workers brushed by the reporter, saying nothing. Newspaper weather reports of the day described it as "fair." But there was nothing fair about what had happened in the minds of the workers. Slowly, they crossed the street to the parking lot, got in their cars, scraped the cinders off their shoes, and drove off. The bars in town, which had always been good places to express anger and blow off steam after work, did a brisk business that afternoon.

I asked Mike O'Donnell what he remembered of the Guterl days. He was one of the 123 union workers still on the job operating an annealing furnace.

"Right away," he said, "when Guterl took over, the guys who worked there noticed certain things. They spent money on their offices; they bought a big bus and went on trips and sporting events. They had rooms at the Best Western on Transit and had parties, but they didn't do anything inside the plant. When the executives moved into town they bought big expensive homes, boats and lived high. When they got the government-backed loan with the help of our local politician, John LaFalce, they spent some of it on the plant, but not enough to keep things going."

"Didn't they buy the AOD, or was it the VIM?" I asked.

"You're right. It was the VIM, but they were not paying their bills and the creditors caught up with them, and they had to file Chapter 11. The banks took over then. But before it actually happened, some of us suspected it was coming. Work orders were canceled, we were asked to hold off cashing checks, and the company suggested a pay cut. A lot of guys didn't take it seriously; thought it was just a ploy."

One of the bills in arrears was for electricity. I thought it an irony that Dan Simonds had moved the plant from Chicago to Lockport to take advantage of cheap electricity. But during the Guterl years, sources told me the monthly electric bill had risen to 1.3 million dollars, a cost that would have been reasonable if sales had kept pace. The plant closed for good in May 1983. I asked Mike what was going on in the seven months between the August bankruptcy filing and the plant closure.

"They gradually reduced the workers over that time," he said. "For most, the day after the bankruptcy filing was their last day. But there was a lot of valuable material in the plant, including steel that had been poured. No more steel was made after the bankruptcy, but the banks wanted to keep people working to get as much money as they could from the assets that were there. One of the union reps said, 'Don't do it. Shut it down. Why let them make money off of us?' But work was work."

Some of that steel, I learned, was the poor-quality steel that had been rejected by customers and returned. It looked good to inspectors who checked inventory when the banks did their due diligence, but it was "garbage" in the words of one of the retirees. It had to be rolled, remelted, finished and re-shipped.

"After the auction," Mike continued, "people were able to come in and purchase mills and other equipment. Some outfits had been interested in buying the plant. We were hoping one of them would buy it so we could keep our jobs. I remember talking to a man from Tuxedo, New York, who was very interested; but he just couldn't raise the money."

A series of banner headlines in the *Union-Sun & Journal* alternately raised and lowered hopes through March and April, 1983. "Guterl Sale Imminent" was a front-page story about the three-way negotiations between the Shelter Rock Investors group of Tuxedo, New York, Guterl, and USW Local 2857. It had been reported in more than one edition in March. Readers of the articles saw it as a ray of hope shining through clouds on a rainy Lockport afternoon. A premature article entitled "Guterl Plant Is Acquired"raised hopes even further. However, those hopes disappeared when attempts by company president, Pinner, and union president, Geier, to keep the plant operating while the Shelter Rock-secured financing failed to take place. "Guterl Suspending Operation" followed by "Guterl Steel Closes as Union Deal Fails," greeted readers in April.

There were other prospective buyers. One of them, Roblin Industries, was local, but they, too, lacked the capital and financing. Precision Rolled Products of Reno, Nevada, had an interesting tie to Lockport. Its roots, in fact, were former Simonds employees. They and a few others would become major bidders at auction.

"What about the workers' pension and benefits?" I asked.

"You know," he said, "big steel had a better deal, full pension and benefits after 20 years. I had 28 years in, but for our contract I needed

30. We had a concession clause that stipulated that if the plant was to be purchased within a year, the existing contract had to be honored. I believe Allegheny-Ludlum knew this and waited more than a year to assume ownership."

"So they didn't have to deal with the legacy costs?" I asked.

"Yes. This is what I was told: the outfit filed in Delaware, which absolved them of any liability, including medical benefits." Delaware was as good a place as any in the country to set up a corporation to shelter assets and avoid tax and other obligations. "The big guys got out without losing anything—the little guys lost their jobs."

When Guterl filed for bankruptcy protection, they had accumulated a 6-million-dollar debt and had run out of cash. Nevertheless, they continued their management style, which included, among other things, deception. While they still ran things, they petitioned the court for money to purchase nickel to complete orders for steel alloys that they claimed they had before the bankruptcy. The court granted the request and they purchased nickel. When the trucks arrived the metal was off-loaded. But instead of making the steel, they sold the nickel, reloaded the trucks and shipped it out. Who got the cash no one knows. Before the bankruptcy, they had also asked the workers for concessions, such as giving up vacation pay. The union grudgingly granted the concession in order to help keep the company solvent. Meanwhile, they continued to pay executive bonuses.

But when it was time to make good on an agreement with the union to share the cost of medical coverage for the remaining workers with Blue Cross and Blue Shield of Western New York, they failed to come through. They had apparently run out of cash, and when the insurance company demanded back payments from January through April amounting to $250,000, worker coverage ended.

Mike said, "They [the workers] had to go out and buy their own Blue Cross and Blue Shield, and it was expensive, several hundred dollars a month, even for a bare-bones policy." The self-payment concession had been part of a pact between management and the union to cover a period of ongoing negotiations with prospective buyers. The newspaper headline on May 12th: "Guterl Pact OK'd, Plant to Re-open," was another disappointment.

Benefit concessions and contract givebacks had become common as the 1980s began. Employer-provided health care had come about almost accidentally as a by-product of World War II wage and price controls; but it had become entrenched, and the workers and their families had come to depend on continuing coverage.

When layoffs began, most thought it was just another bump in the road that they would pass over, as they had done so many times in the past. But even the union workers at the major steel producers suffered. Bethlehem severely underfunded their pension, as well as their health and insurance program with the Massachusetts Mutual Insurance Company. The federal government took on their underfunded pension costs, but the pensions were not as generous as workers hoped they would be. The men over sixty-five got most, but not all of their pensions. The younger men got much less. Mike O'Donnell was one of those men who started collecting a reduced pension at age sixty-two. The same situation held for company-sponsored health benefits and insurance. The retired men who were over sixty-five had Medicare, but the younger men had to buy their own insurance. I asked Louie Koel, who had participated in the negotiation of the 1980 contract, about the deal.

"I feel real bad about it," he said. "I was a part of that union. I spent sixteen years as an official, three terms as treasurer, legislative rep, and grievance committee man. My Simonds pension check is only half of the amount that I get from Virginia Tech, and I worked there a lot less years." Louie had found employment locally after Simonds, but he eventually was hired in the athletic department at the Blacksburg, Virginia

University, where he had family in the community. "Our committee had a list from meeting with the rank and file," he said. "The company had their position and we went over one item at a time. You get one and you give one, you don't get everything you want, but I'm sorry we didn't do more to protect our benefits."

"At the time of the bankruptcy filing I read that the union was caught off-guard," I said. "Your president didn't know what Chapter 11 involved and what it meant for the union workers. Would wage and benefit concessions have helped?"

"I don't think so at that point," he said. "The union had already made concessions including pay cuts and tried to work with management to keep the company going. Even after the bankruptcy, the months of uncertainty, the rumors of possible buyers, and pressure from his coworkers, he [the president] worked with the city leaders and the new owners trying to convince them to rehire the steelworkers who had spent their lives at the mill. You couldn't fault him for lack of trying."

But some of the workers did. Mike O'Donnell told me several union guys would have bent over backwards, but others did not want to take any more pay cuts, hoping to hold onto their $25-an-hour jobs and middle-class security. "We didn't stop to think about where we would find a job with similar wages," he said. "When the reality of what had happened started to sink in, it was very depressing for the guys."

A 2017 Pulitzer-winning, Tony-nominated Broadway play, *Sweat*, intensely dramatized a group of devastated steelworkers, some manifesting fear and stress; others disbelief and anger at having lost their union jobs with good pay and benefits—the source of their hopes and dreams. The stage setting, a bar in Reading, Pennsylvania, could have been the bar room at Little George's, where the plant closing and future prospects were the subject of lamentations for a long time afterwards; as I suspect it was the same in bars wherever steel mills had closed and were eventually shuttered. J. D. Vance's book *Hillbilly Elegy* about working-class loss in a Rust Belt town, and Rick Campbell's poem, "History of Steel"—which reads like an elegy to idled steelworkers—are recent literary echoes of that period.

Labor unions have taken a beating since 1980, the struggles of early twentieth-century workers for justice and legitimacy largely forgotten. Each Labor Day provides the opportunity for newspaper columnists and op-ed writers to review the history of the labor movement and offer opinions on the pros and cons of unions and suggestions for improvement. Opinions vary widely, but there seems to be little doubt that the cumulative effects of various political trends, regulatory policies, and judicial rulings have tilted bargaining power away from the worker and unions and toward employers and corporations.

New York has always been a strong union state, but in recent years unions have been accused of doing less to protect workers and more to enrich themselves. Membership has severely declined as industrial workers have gained freedom not to join; wage concessions and givebacks have regularly occurred, and unions have not been able to influence management to make good on promises to retrain workers whose jobs became vulnerable to automation. Some studies have shown that this decline in union membership has hurt all workers by contributing to wage stagnation and financial inequality; in fact, the decline of the middle-class life style is stated to correlate with the decline of unionized work. Interestingly, as this has occurred in the manufacturing sector, health care industries and academic institutions have seen growth in professional trade unions. Service employees have also begun to seek collective bargaining for better wages and working conditions.

When labor unions began to face big challenges in the 1980s, a former AFL-CIO president was often quoted: "We're not interested in a society where McDonald's employs more people than U.S. Steel." However, that is what has happened. Many more people make hamburgers and serve coffee in this country than turn out steel product. So it would seem appropriate that the recent series of protests and strikes in fast food has launched a drive for unionization of the burger assembly line.

■   ■   ■

I asked Lou Valery what ended his career at Guterl. He told me that he had had an argument one evening over a batch of steel with Guterl's quality control manager. The quality of the steel made at Simonds had never been an issue, but Lou felt that some of it was becoming less than optimal in the early '80s. One evening he objected to an order that he believed did not meet the customer's specifications. The dispute became heated. "Hey, you do your job and I'll do mine!" the QC manager shouted. The interchange got the superintendent's attention. When he had sorted things out, he directed that the job be redone and that Lou personally deliver the steel to Jessop Steel in Pittsburgh for custom cutting and shipping to the Italian customer who had ordered it. While at Jessop, Lou saw an idle sheet mill and asked if it were for sale. It was. A week later, having given their notices, he and Hemphill drove to Pennsylvania. They spent a couple of weeks together tearing out the mill. Through eighteen-hour days, they covered "each other's backs," exhorting one another.

After the equipment was trucked back to Western New York, they spent more grueling hours setting up the mill in a rented building in nearby Akron, a former Carborundum facility. They often found themselves at important crossroads with decisions regarding hiring, finding potential customers, and startup costs. When the public auction of the Guterl plant and equipment was held, they purchased individual pieces of equipment, including the sheet mill where my father had spent so many years of his life. My father had been retired for several years at that time. He had come to terms emotionally with the demise of the mill and was happy to learn that his sheet mill had been rescued from the trash heap.

Six of the production workers the partners had known and liked at Simonds, including rollers, roughers and finishers, a heater, a blacksmith, and the crane man from Lou's sheet mill department were initial hires. Dave Craine, who was capping twenty years at Simonds, including stints on the 10-inch bar mill, band mill, and crane, was hired into the shipping department. Dave also had acquired a commercial truck driver's license. Among his assigned jobs at Niagara was trucking steel from mills in Ohio, Pennsylvania, and Welland, Ontario, back to Akron.

Valery and Hemphill began their venture undercapitalized. They needed to borrow more money from the banks than they were willing to risk at that time, though the partners disagree on the extent of the capital shortage. Hemphill had shopped his business plan to various banks, finally getting a loan from Marine Midland. "You need more money," the loan officer said, "but we didn't take it," Lou told me. It took about eight months to get going with their first production. "The first year was rough," he said. "We didn't take salary for months, but then we got busy in the second year and business was good from then on."

Their vision was to occupy a market niche wherein the company did not make steel as they did at Simonds, but rather purchase the specific tool or stainless steel alloy a customer wanted from other steelmakers and then roll it to their specification, quickly and efficiently. Customization and speed, in Lou's opinion, were the keys to success. Everyone, it seemed, had discovered that ordering steel overseas from Japanese, South Korean, and European producers was cheaper. But Lou was able to convince potential customers that he could customize the product size and volume, deliver it faster, and step in when foreign steel producers ran out of their product. They never competed on price, but on service. "That was the best part," he said about the marketing strategy. "It worked."

I asked how it all ended for him. "I got tired of working." He persevered for twelve more years, and then Hemphill bought him out in 1993, and he retired to sunny Florida to play golf far from the less-than-optimal weather of the Western New York. "From the beginning, I knew the business would make it," he said. "Even when the money was low, I knew the bank would give us what we needed."

Hemphill successfully continued the business and then transferred ownership of the company to employees through an ESOP (Employee Stock Ownership Plan). The arrangement preserved the company's culture and place in the community, avoided layoffs, further incentivized the employees, and allowed Hemphill to cash out. He remains chairman of the board. Niagara Specialty Metals today is still doing well as a fully employee-owned company.

Another entrepreneurial success story by former Simonds employees occurred in 1978, after Guterl took over the Simonds Division of Wallace Murray and five years before Valery's departure. That year, Harold F. Kinsler joined two Simonds metallurgists, Dick Cook and Don Gatsby, in a risk-taking venture that ultimately provided considerable profit as well as paychecks for many workers. Gatsby was a big-gamble, big-profit kind of guy; and Cook, who had noticed an available building while on a trip to Reno, Nevada, for a metallurgist meeting, followed the old motivational impetus to "go west." Kinsler shunned a significant financial offer to remain at Guterl, and with his partners and a few others, traveled to Huntington, West Virginia, to tear down a 10-inch bar mill they had purchased from a shuttered steel factory. They trucked it to Reno, set it up and started rolling high-temperature, titanium alloy purchased from an Oregon supplier. It took a year to start making money, but by the third year the company earned 10 million dollars and was listed among the Fortune 500 companies. That was the year Guterl was up for auction, and Gatsby wanted to spend most of the $10 million to buy it. However, he ended his bidding somewhere in the $8 million range, and the asset went to Allegheny Ludlum. "I'm glad we didn't get it," Harold Kinsler told me. "We probably wouldn't have survived back then." Precision Rolled Products lost that bid, but continued to grow. It flourishes today in Reno, Nevada, one of the better cities in America for startups, and where Harold Kinsler is happily retired.

■ ■ ■

It was getting to be late morning, and Michael and I had seen a lot of rusty walls, doors and moldering equipment. I thought it might be time to leave. We had moved from the sheet mill area to the inspection area, now largely devoid of the presses, punches and shears, just some debris on the ground that crunched when we stepped on it. I recognized a residual oven from my working days near a tarnished corrugated wall. It had a rectangular rusted frame on a roller base open at both ends. Steel could be heated and softened by passing it from one end to the other. My father and Mr. Ciarfella often broiled steaks on that oven when they worked weekends cleaning the mill pits. "Time for one more photo," I said. Michael shot the oven, then led the way out through a rusting doorway. We retraced our way through the weeds back to the gap in the chain-link fence where we had squeezed in earlier that morning. The muted sounds of work emanating from a gray building at the west end of the property reached us as we walked to the car, relieved that no one had spotted us.

There were a lot of things on my mind when I started out in the morning. The nostalgia was not completely fulfilled, but I enjoyed sharing the experience with my son. I had found answers to some questions, and the visit had afforded me the opportunity to again reflect on the meaning and value of the hard work and perseverance I had found at Simonds. It reinforced the work ethic instilled by my parents, which had served me well on my journey out of Lockport.

An autopsy has many purposes besides uncovering the cause of death and the events leading up to it: the pathogenesis. It may find previously unknown illnesses with genetic and public health implications, identify ways to improve care; and perhaps most importantly, it can bring closure to a family, such as the industrial family at Simonds. I had what I needed and I hoped others would too. The time had come to sew up this body of work, leave the morgue and write the report.

"The past isn't dead. It's not even past."
—William Faulkner

EPILOGUE

The original Simonds-Wallace Murray-Guterl operation no longer exits, but a legacy of steelmaking remains at the site under the Allegheny Technologies Inc. (ATI) banner, the sole remaining steelmaker in Western New York; and though not connected to the original Simonds family, the Simonds name lives on as Simonds International in Fitchburg.

The Wallace Murray Corporation was purchased by Household International of Prospect, Illinois in 1981, with Wallace Murray and Simonds becoming a subsidiary of Household's manufacturing arm. Seven years later, senior management and private investors purchased the Simonds division, returning it to private ownership. (The original Simonds stock was first listed on the New York Stock Exchange at $35 per share in 1937. Although it was publicly traded, most of it was held by the Simonds family.) The new company, named Simonds Industries, acquired a number of saw blade and knife companies in the United States, Canada and England over the following decade. The ghost of

Abel Simonds and his descendants hovered on Halloween—October 31, 2003—as Simonds Industries merged with International Knife & Saw (IKS) to become Simonds International, currently the major sawmill product manufacturer in North America, and an international player in the metal-cutting bandsaw market.

*   *   *

Jim Calos was one of the salaried employees retained after the remaining work force was terminated and the Guterl plant closed in March 1983. Jim and Burt Malcom had both recently retired from Allegheny—Jim as the melt shop superintendent after a thirty-seven-year-career that began with Guterl, and Burt as a production manager after a fifty-three-year-career dating back to Simonds Saw and Steel. Michael and I met with Burt and Jim over coffee one morning in Lockport to discuss Allegheny operations since 1984, changes in the steelmaking process from the earlier Simonds era, and what to expect from a tour of the ATI facility that Jim had arranged. Both men had given me considerable information over the phone and in emails, and they had provided catalogs, photographs, and newspaper articles from the Simonds era on an earlier visit.

Jim, a Lockport native and graduate of DeSales High School and the University at Buffalo, had worked as an estimator and expeditor for a stainless and tool steel distributor in Western New York. In 1978, he was hired at Guterl in the same capacity, and the following year he became their product manager for tool steel. When the plant closed, he was among those charged with maintaining the plant's security and equipment and liquidating the inventory. "We took our sweet time liquidating the material," Jim said. He went on to suggest I talk to Reggie Buri, who he said represented the bankruptcy bank in the onsite discussions with Allegheny Ludlum and was present at the bankruptcy auction in Pittsburgh.

Reggie Buri was a Lockport High School graduate who had gone on to the General Motors Institute in Flint, Michigan, and a career as a product engineer at Harrison's. In 1978, he joined Guterl as chief of maintenance. Following the bankruptcy he was retained by the Department of the Interior, the government agency that had guaranteed the $15 million Guterl loan. The Consarc Corporation was the intermediary chosen to recoup the loan through the federal bankruptcy court. Buri confirmed Consarc charged him with maintaining the plant and its assets while conducting interviews with prospective buyers. He oversaw the efforts of a few other men who maintained the building's equipment and inventory, provided security at the gate, and made rounds and inspections throughout the plant.

Theft had become a security issue after the plant closed. Valuable material disappeared. Some thieves were brazen enough to cut a hole wide enough in the chain-link fence to drive in pickup trucks and make off with copper, steel, and other materials.

Several companies were interested in the melt shop, specifically Guterl's clean steel technology equipment, especially the VIM and ESR units, which had only been subjected to eighty heats. Buri established himself in the office building and entertained about three dozen prospective buyers. Allegheny Ludlum Steel of Pittsburgh, Precision Rolled Products of Reno, Nevada, and Curtiss-Wright of Wood Ridge, New Jersey, were the last firms standing at the U.S. Bankruptcy Court auction held in Pittsburgh in August, 1984. Allegheny won the bid with a 9.5-million-dollar offer. When the pre-purchase review was being conducted in Lockport with the Allegheny people, the president of Allegheny, Richard Simmons, got wind of the radiation problem and called Buri from Pittsburgh about what he considered to be the theretofore-undisclosed radioactive contamination. Buri told Simmons that the issue was "not a secret locally," and followed up his conversation by sending Simmons *Union-Sun & Journal* accounts and a previously printed brochure listing the Department of Energy testing results.

Allegheny wanted the VIM more than any other asset, so Simmons agreed to honor the sale, but refused to take responsibility for the contaminated area, which included two portions on the Simonds site: several buildings and a landfill. When the federal bankruptcy court judge approved the exclusion, the 8.6-acre section was fenced off as the "excised property" and today remains in the hands of the United States government's Army Corps of Engineers. A sign on the chain-link fence along Ohio Street marks the line of demarcation between the excised property and Allegheny Ludlum property. The ultimate responsibility and time for the site's cleanup are unknown.

The corporate leadership at Allegheny, the country's eighth largest steelmaker and leading stainless steel producer, had decided that their Lockport plant would focus solely on super alloys for aerospace. On October 14, 1984, its Allvac division—Allegheny Technologies Industries (ATI)—began operations in the Lockport plant. It was an inauspicious start. An indoor ribbon-cutting ceremony attended by the Allegheny president and a group of his executives included Lockport's mayor, Tom Rotondo, a former Simonds steelworker in the mid-50s, and other elected city officials. Meanwhile, forty former Guterl union workers picketed outside. They were protesting Allegheny's non-union shop strategy and hiring policy based on interviews that sought certain skills, aptitude tests and a high school diploma. Seniority and steelworking experience alone were insufficient. Pain and disappointment on the picket line were palpable for anyone who was there. The mayor and city council had attempted to intervene on behalf of the former workers, threatening to withhold a $740,000 Urban Development Action Grant. But Simmons was adamant and threatened to pull out of the deal if Allegheny could not implement its own management and work practices. In regard to the union, the president kept his cards close to his vest, but allowed that he would be open to negotiations "down the road."

It had been a bumpy road over the previous two years with layoffs, plant closings, and dashed expectations, but finally a reopening and a new beginning was anticipated. The city was glad that steel was being made again, even as hopes for jobs for most ex-Guterl workers vanished.

Allegheny did rehire eighteen former workers who had the knowledge and experience in running equipment and processes that Allegheny lacked, and they made those hiring reasons quite clear to everyone. Reggie Buri, Jim Calos, Burt Malcolm, Gordon Martin, Bill Parsons, and Tom Nicholas were among those hired into salaried positions based on experience they had acquired in melting, remelting, cold rolling operations and equipment maintenance; while Mike O'Donnell was one of the few former Guterl employees rehired into their non-union work force along with men who had no steelworking experience. A bi-panel photo in the *Niagara Falls Gazette* on the day ATI began operations shows an operator at the controls of the vacuum induction furnace next to a picketing worker outside the plant holding a sign, *"30 years' experience; I'm not qualified?"*

Mike began, quietly at first, to reorganize a union while Allegheny tried their best to discourage his efforts. Local 2857 had ceased to exist with the Guterl bankruptcy. Initially, Jim Geier, the last union president, had been kept on with Buri's crew because at the time no one knew whether the plant would reopen as a union or non-union shop. When Allegheny took over and Simmons announced the plant would start out as a non-union shop, Geier was history.

All of the union records had been lost along with the personnel records from the early days of Simonds. They were part of the piles of paper that had been unceremoniously dumped like trash in the cast magnet area after the plant's closure. In his organizing efforts, Mike got some help from union workers at the Allegheny plant in Pittsburgh, who surreptitiously shipped him helpful information in supply boxes, including their contract, wages and benefits package. It took two years of travel "down the road." They narrowly lost the first vote, but Mike and his fellow workers were happy that their voices were being heard, and they had gotten the opportunity to organize. On their second attempt they reformed Local 2857, and Mike was elected president.

ATI began manufacturing alloy steel for jet engines. Their initial intention was to use the EAF and AOD units along with the vacuum induction melting process. However, it became apparent that ultraclean steel was becoming an absolute requirement for aerospace in contrast to the less-vigorous requirements for automobile products. Since that goal could not be met with the EAF and AOD, a decision was made to shut them down in favor of vacuum induction melting (VIM) and electric slag re-melting (ESR). Vacuum arc re-melting (VAR) units were subsequently added to the operation. Jim Calos had become the melt shop superintendent for the VIM in 1987. He explained that steel made in the normal industrial atmosphere acquires the contaminants in that atmosphere. The VIM provided the controlled environment to remove impurities while converting raw nickel and more exotic metals, such as titanium and other oxidizable metals, into super alloys. But for many applications, the VIM alone could not make the alloys "super." Even trace amount of oxides, lead, bismuth, and other elements in the ingot had to be more completely removed. The re-melting of the ingot as a consumable electrode (the steel electrode itself is melted) in the electro-slag re-melting (ESR) unit could make the alloys cleaner, but adding vacuum arc re-melting (VAR) was the process that put the "super" into super alloys. Think about the water we consume nowadays. Perfectly clean and drinkable after it leaves a modern water treatment plant, we further subject it at home (or buy it) to reverse osmosis and ion exchange resins to make it ultra-pure for discriminating tastes. Running distilled water through a Brita filter might be another analogy. Today, ATI uses specific types of clean scrap metal known as "revert" in both of its remelting processes for the high-quality ingots it produces. It's a far cry from the scrap metal I saw in the trash heap at the Simonds yard in the 1940s.

ATI occupies a particular niche in the specialty steel industry, along with a few other companies that use VIM furnaces. "Thank God Guterl used some of that 15 million dollars to buy the VIM," Jim Calos said. "That's why ATI is here today." The question is, will they continue to make steel here? Who would have thought that the major activity at Kodak Park, now Eastman Business Park in nearby Rochester, New York, where once thousands of employees made and sold most of the world's photographic film and paper, would not be Kodak, but rather a variety of leased business operations, including a bottler of pasta sauce?

The day after Michael and I met with Jim and Burt in the morning and a group of retired steelworkers at the American Legion Post in the afternoon, we drove to the ATI Specialty Material's plant for a tour that Jim Calos had arranged for us. It was a sunny spring morning. The buds on the heavily tree-lined streets and along the canal were beginning to leaf out. It was the kind of May morning I fondly remembered in Lockport, with flower beds of daffodils tilting their yellow heads to the sun; the irises and peonies shooting stalks skyward, portending a display of gorgeous blooms in an array of colors. We pulled into the drive by the *ATI-Lockport Operations* sign, the same sign that had once read *Guterl* and previously, *Simonds Division of Wallace Murray*, and before that *Simonds Saw and Steel*, and parked by the brick office building. As we walked to the security kiosk, the soft rumble of work emanated from a tall gray rectangular building on the west end of the property adjacent to the "excised area" of the old mill.

Our tour guide, Tom Wager, shortly arrived at security in his hard hat and work boots and escorted us to the office of his boss, the newly named plant manager, Craig Rowmanoski. The plant had recently restarted operations after a six-month lockout that involved not only steelworkers at the Lockport plant, but also 2,200 other ATI workers in twelve states, including 1,500 in Pennsylvania. Newspaper accounts suggested that management and union workers were satisfied with the new four-year contract, which protected retirement benefits and jobs from outside contractors, maintained the grievance process, and installed a new profit-sharing program. Both Tom and Craig seemed optimistic for the future.

As we chatted in preparation for the tour, we were presented with boxes of archival material from the Simonds, Wallace Murray, and Guterl years. Among the old catalogs, newspaper clippings, drawings, and photographs were packets of black-and-white 4"x5" medium-format negatives. I pulled one of them from the packet and held it up to the light. I was stunned to see the image of a worker tossing a shovel of metal into the door of the old electric furnace as smoke billowed out between him and a fellow worker. Excitedly, I continued to pull and examine negative after negative showing a tap and pour into a pit ladle in the old melt shop, the ingot molds, and men at work on the hammer, bar and sheet mills. The box contained Kodak Safety Film color negatives as well, including one of a tap and pour from the electric furnace in the new melt shop. I held images I had spent months trying to find, images of processes that I struggled for words to describe. In my hands the old adage "a picture is worth a thousand words" had literal meaning.

After Michael and I were fitted with hard hats, orange vests and work boots, Tom walked us out of the office building toward the ATI building. On the way, we stopped at the metallurgical lab, where the manager offered to test the ring I was wearing. My grandfather, Antonio, had it made for me in 1955 when he was working in the bar mill finishing department. I didn't know much about it except that he told me it was made of stainless steel. The test showed that it was stainless "304," the non-magnetic, 18–8 percent chromium-nickel alloy, a nice thing to know, even sixty years later.

Actually, my grandfather made more than rings. The drawers in his kitchen cabinets had several knives that he made on his lunch breaks from scraps of tool and stainless steel. He also made some of them for my mother. In fact, I don't remember any store-bought knives, but I do remember the hours he spent honing his knives on a whetstone on his basement workbench. I was told that one worker was such an outstanding knife craftsmen that he was able to make and sell hunting and jack knives to Simonds employees and others in the community. Gordon Martin showed me a well-crafted knife at his home that a steelworker had made for him. A recent article in the *Union-Sun & Journal* featured a former steelworker, Roger Bil, now eighty-six years old, who makes knives in his home workshop. It was a craft he had acquired on his free time while working at Simonds shoveling scrap in the melt shop in the 1950s, a hobby he continued well past retirement. My grandfather also made jackknives, lethal-looking, sword-length barbeque skewers, and a Jew's harp, a small lyre-shaped steel instrument that he held between his teeth and twanged notes through inspiratory and expiratory breaths. I still have those items. He also fashioned an ingenious spout for his wine barrels with a special "key" he always kept in his pocket. He attempted to make straight razors, although he was never able to bring his blades up to the quality of those in the German razors he much admired.

Leaving the lab, we passed along the fenced-in corrugated-metal building that had once housed the Simonds rolling mills which Michael had explored earlier. Broken windows, torn and missing panels, and an open doorway afforded glimpses of the interior, but the fence's padlocked gates prevented us from entering. However, we were able to enter at the end of that building through a gate that Tom unlocked for us. It was Allegheny Ludlum property, but the south end of the interior had been walled off from the excised zone. This Allegheny property was now used principally for storage, but it had once housed Simonds' bar mill inspection and shipping department. Black-and-white photographs I had just examined in the ATI office revealed this area crowded with stacks of steel bars, sheets, and rows of steel strip coils. Men, inspectors, and machinists were scattered among the steel products, working at presses, planers, and lathes while others stood by high-speed milling machines. I could picture my grandfather among them making my ring. The shipping area resembled a garden of steel with stacks of bars, sheets and coils waiting to be loaded onto boxcars for distribution to Fitchburg and other customers. In those photographs of the boom years, the building seemed a hive of activity.

But it was silent now. Daylight from the bank of windows high on the eastern wall illuminated the girded ceiling with rows of suspended unlit lamps and fell upon stacks of giant, grainy-brown iron ladles, massive hooks, and other mothballed equipment. We maneuvered around, following Tom, listening to his explanations that faintly echoed while I looked for a possible entrance into the excised zone. I knew just beyond the wall was the site of the old Kinsler restaurant and beyond that the 16-inch bar mill. I wasn't anxious to re-enter that space, but I knew Michael was always looking for an opportunity to improve on the photos he had previously taken with a fresh perspective. I was unable to find access, so after taking some photos of the old ladles, we left the building to continue the tour, passing in-ground portals the Army Corps of Engineers had placed to periodically sample the groundwater. When we finally entered the ATI building, I was struck by how different the environment was from my memory of Simonds. Burt Malcolm suggested during the interview the day before that I would be walking into a completely different world than the one I had experienced. He was right, and Reggie Buri implied the same thing when he said, "My career in steel was like living two life times, the first [Guterl], old school, low tech, and the second [ATI], new school, high tech."

Tom Wager had us enter through marked doorways and walk along yellow-striped paths, so unlike the pre-OSHA environment of Simonds. We encountered relatively few men at work and seemingly little activity in the middle of the day. Machines and computers had replaced hands-on manpower. It was much quieter, not the loud pounding noises that I remembered from the '50s.

ATI melted specially processed scrap metal with alloys to form ingots, which they re-melted and reformed as high-purity aerospace steel, but that was it. The end of the line at this plant was making steel ingots.

Just past the entrance we paused to watch a worker who stood near the giant Consarc re-melting machine. Using a hand-directed crane, he eased the hook and chain holding a newly re-melted ingot to the floor. Then we proceeded past other large gray steel cylinders lying side by side. Some of them still radiated heat. The ingots would be inspected, stamped and shipped elsewhere, where they would be finished into the forms that would be machined for jet engine components: fan blades, compressors, combustion chambers, and turbines, each of which required a different composition of alloys.

Gone were the presses, rolling mills and finishing operations of the earlier Simonds era, with all the sights and sounds of men at work. Up on the high platform of the VIM, I peered through a glass portal and saw molten red steel surrounded by a wreath of gray slag. The porthole provided the only evidence, to my eye at least, that steel was being made in that building. The metal, which was destined to become a solid ingot and an electrode to be re-melted and recast, was not so different from what it would have looked like in any of the three electric furnaces that were now just museum pieces silently perched on their pouring platforms.

The mini mills of America, including the specialty steel segment, have been successful; but, as steelmakers of the nineteenth century realized, steel production involves great energy cost. In the twenty-first century Nucor and specialty mill operators, not having access to hydroelectric power, have relied instead on relatively inexpensive and abundant, home-grown energy in the form of natural gas to make steel manufacture profitable. But the future of cheap natural gas is uncertain, and foreign competition has been eroding Nucor's $20-billion annual sales success and that of other steel producers as well.

The steel industry has been reeling from a plunge in steel prices as lower-cost steel from Asia, Canada and Mexico has flowed into the United States. At the beginning of the twenty-first century, steel production by the U.S. and China was roughly equivalent; but with heavy government subsidies, China's output increased ten-fold, while U.S. production decreased. In 2016 steelworkers from across Europe met in Brussels to protest China's dumping of steel on the market at below-market price. In an effort to combat the dumping, steel company leaders in the United States have resurrected efforts of earlier decades, asking the Commerce Department to impose heavy tariffs, quotas, or both, on foreign steel to stave off another round of layoffs and bankruptcies.

Before the passage of the income tax in 1913 (13th Amendment), tariffs and duties were how the U.S. government obtained most of its revenue. Tariffs were considered a good strategy; Carnegie himself had conceded that this indirect subsidy had benefited him and other manufacturers. Why not try it again, even though history had shown the strategy sometimes backfired?

John Ferriola of Nucor was one of a dozen steel and aluminum executives called to the White House in February 2018 to discuss this issue. Tariffs and quotas serve the same purpose, protecting domestic industry—tariffs mainly through pricing, and quotas by restricting quantity. Economists site advantages and disadvantages for both. Many believe tariffs are controversial, generally don't work, and potentially unleash trade wars and lingering animosity. The price of protectionism in the past has been increased costs to domestic manufacturers and consumers. The effects of tariffs within the steel industry, downstream users, and their customers vary with the geography of competition for a particular steel mill. Some steel producers of structural beam and rebar view their rivals in Canada and Mexico with a jaundiced eye, while the stainless, tool, and other alloy producers see their threats in Asia and Europe, or here at home.

American automobiles, for example, require sheet steel for bodies, structural steel for frames, and stainless and other special alloy types for engines and components. The aeronautics and space industries require even higher-grade alloys for its jet engines. Most of these products are preferentially supplied by American steelmakers, so tariffs and quotas may not affect how these companies face off against each other. However, while buyers may prefer U.S steel, they will only tolerate a limited price increase. Steel producers in Lockport and Akron, New York, may see it one way; steel producers in other cities, another. At the heart of this matter is the local economy in the particular steelmaker's community—the consumer and its nearly forgotten steelworker. In any case, there promises to be winners and losers as well as unintended consequences in any quota and tariff strategy.

One thing does seem certain, however: The global playing field is often unfair, and some workers suffer from trade policies which are forged in a geopolitical context. But foreign steel competition is not the only factor. The steel industry is impacted by new technology, events in other industries such as oil and gas, and currency manipulations. Changes in these sectors, reflected in lower commodity prices and declines in the economy in general, can reduce the demand for steel products. The dangers that lurk in the global economy can't always be avoided; there are things about which little can be done, even with foresight in recognizing threats, adapting to change, cutting costs, and improving productivity. A former Kodak engineer told me the company saw digital coming: "They just didn't see this," he said, as he pulled his iPhone from his pocket. They made nearly all of their profit from film and paper, not cameras. He claimed the company's position was, "You can have our cameras, just use our film."

Different companies may need different strategies. Perhaps Kodak needed to leave photography and transition into an entirely different industry. They took a shot at pharmaceuticals, but that didn't work. Kodak's current vision is to become a site for innovative technologies. Who knows? The Fuji Corporation, which had its roots in photographic film, was able to innovate new products and services such as medical systems. As I was writing about the crisis in steel in an early draft of this manuscript, I listened to a story on public radio about the same crisis facing aluminum industry towns and workers. It was, as Yogi Berra has been quoted, "It's like déjà vu, all over again."

In the second decade of the twenty-first century, articles began to appear in the popular press about the future of work. *Washington Post* columnist Michael Gerson wrote that whole categories of labor have been replaced by new technology. "A way of life in which increased productivity resulting in higher wages and a realistic shot at economic advancement is failing." He added that the effective collapse of the blue-collar economy has threatened communities and placed many of them in permanent recession. Educating and equipping workers for this more-technical economy, where artificial intelligence and robotic manufacturing play a growing role, is the challenge as he and others see it. But none of this is new. Four decades ago, editorialists opined that the labor force needed training and skills for the newly emerging technology in diverse industrial fields.

Manufacturing may return in some form, but with new technology and methods replacing lower-skilled 1950s-era manpower. Every job is likely to require new technical knowledge, which will in turn require new educational strategies for students of all ages. However, a real concern is that the advance of technology is so rapid it threatens to outpace the time required for many to be educated and adapt. Customized manual rolling operations at specialty mills seem safe, but automating certain component activities may become practical and cost-effective in the near future. Automation with robotics may save steelmaking in this country, but at the cost of fewer workers with higher skills. The downward trajectory of job numbers in all industries and the upward trajectory of population have prompted some futurists to speculate about a universal basic income.

I asked Harold Kinsler if he thought worker positions at Precision Rolling would be lost to automated robots. I pictured robots rolling a sheet of steel, pitching and catching it under the direction of a skilled worker who programmed, monitored and maintained the robots.

"No," he said, "and you want to know why? We did small custom jobs. It's too expensive to automate small tonnage custom work. It may be cost effective to automate similar day-to-day processes that were repetitive and larger-volume, but not the type of work we did."

Lou Valery agrees with him. When I asked him if the potential existed for automation that could make the manual steelworker at Niagara obsolete by replacing some of the activity with robots, he said, "There is no way they can use it to do the kind of jobs we did there. It's not just a matter of rolling, pitching and catching a sheet and extending it, say, from 12 to 24 inches. It has to be brought back and forth from the annealing furnace and then turned this way and that several times. I don't know how a robot could have the judgment and manual dexterity to do that, but go take a look for yourself and decide."

I did. On a late September day, I drove out to Niagara Specialty Metals with Dave Craine, who had arranged an interview and tour with Robert Shabala, the company president. Niagara County had changed since I had grown up there in the 1950s, but there were certain images of the ways things had always been: the fields of corn, the orchards of red apples that triggered fond memories. I had forgotten how charming the wooded countryside was with the dense cordon of tall hardwood trees in their early autumn colors encroaching upon the winding two-lane backroads to Akron.

On our tour, Shabala pointed out that the fully employee-owned company had added additional capabilities designed for aerospace and electronics to its primary tool and stainless rolling mill operation, and had developed a working arrangement with several alloy steel producers, including Crucible Industries in Syracuse. Piles of Crucible's steel alloy flats were in and around the buildings we toured.

Peering into one of the buildings on the tour, I could barely make out the men working in a dense fog of smoke from a molybdenum alloy heat. Through the smoke I could see that a forklift had replaced the backbreaking work of leveraging a fiery sheet of steel into a flattener. The highlight for me was finding that the finishing mill where my father had rolled so many sheets of steel was still in active use. I didn't recognize it at first; the housing's gunmetal dark gray had been painted blue and silver. I watched two men pull a red-hot blank from an annealing oven onto a long-handled buggy and wheel it to the mill. Since they bought other Simonds equipment at the auction, it could have been the same buggy I had once used. Just like the old days, the mill hands passed the sheet back and forth through the rollers, but now with the added assistance of lifting tables. I thought about my father and his chronic back problems that those lifting tables might have prevented. If there was one benefit for future steelworkers from automation, it was longer careers without chronic injury. The steel supplied from Crucible now had Space-Age composition, but what I was witnessing was still the craft of handmade steel, something a worker—not a robot—could take pride in. In the pursuit of automated efficiency and precision, what will become of this sort of human craftsmanship—not just in steelmaking, but in all manufacture?

The sheet was flipped and turned and flipped again and then pulled onto a table. Fifty years ago at Simonds I would have stacked those sheets, but now at Niagara, the sheet was steered from one table to another by a shear man where it was cut into sections. "Can't all of this be automated?" I asked Bob Shabala.

"Actually, a producer in Austria does have the front end of this process from the furnace through the rolling mill automated," he said. "Then it becomes a manual operation. Reheating and rerolling is a challenge to automate."

Now, but for how long, I wondered? New technology in all fields is evolving at blinding speed. Personally, I've worried about dire predictions that the microscope, which has been around for 400 years, may become obsolete as a diagnostic instrument, and with it the pathologist who is largely defined by that instrument. Training computers in image analysis, so-called "machine learning," combined with molecular diagnostics, has the potential to make the prediction come true. But like Lou and Harold, I don't see an expiration date any time soon, and I'm not going to sell my microscope just yet.

Hopefully we're right. Otherwise, the photos depicting scenes of blue-collar workers on factory floors, as described and illustrated at Simonds and updated at Niagara, and men seated at microscopes are likely not to be seen again except in the archives of historical societies.

* * *

Simonds' Lockport plant and its successive owners, Wallace Murray and Guterl, rose and fell into ruin in the course of 116 years; a history traced by crucible, electric arc, and vacuum induction steel manufacture. The latter, with its refined re-melting processes and innovations, will hopefully keep steelmaking at ATI in Lockport viable into the foreseeable future. But what will become of the moribund buildings on a radioactive parcel of land by the Erie Canal, one of the "rusted-out factories scattered like tombstones across the landscape of our nation"? The buildings and ground where once generations of dirt- and sweat-soaked men turned scrap metal into alloy steel, poured from a furnace into ingot molds, ground, pressed, and rolled into bars and sheets; finished and shipped to producers who churned out the many products that served our nation's needs in war and peace; now only serve as a symbol of uranium's legacy: the contamination of soil and groundwater and the need for an expensive cleanup.

When companies declare bankruptcy and leave contaminated sites behind, without having posted bonds for cleanup, they can become Superfund sites and a burden to taxpayers. The answer to when the cleanup will occur has been sought by Lockport's citizens, city and county officials, since the federal government allowed Allegheny to excise the property from its purchase more than thirty years ago, and years after the Department of Energy asked the Army Corps of Engineers to add the site to its cleanup program.

As I was working on this manuscript in 2017, an article appeared on the front page of the *Union-Sun & Journal* announcing that the Army Corps of Engineers was expected to issue a proposal for community input for the Simonds site cleanup, with a so-called Record of Decision to be be issued in 2020. The cleanup will entail the demolition and removal of the buildings and excavation of the soil. The radioactive residue in the soil will last for hundreds of years if not removed and disposed of properly. It's been a long time coming, but the extirpation of the decaying factory eyesore on Ohio Street is expected to bring welcome closure to a tumultuous chapter in Lockport's industrial history.

THE HISTORY OF STEEL
—By Rick Campbell

In winter we wore thick coats
And hats with ear muffs folded down,
Snapped under our chins.
We carried black lunch boxes
To the bus stop and waited
In the sharp, wet wind.
Others drove soot-streaked
made in America cars and jammed
traffic three times a day.

When everyone was working overtime
and money was thick as air,
sons started coming home in boxes.
Few saw how high prices had climbed.
Then the lay-offs began.
Union reps mumbled patience
and it was easy enough to take.

Benefits were good. Money
that had come so fast, before
anyone learned how to spend it,
sat abstract in the bank.
People said we were doing ok.
Had time to panel the basement,
go fishing up at the lake.

History was right on top of us,
too close to see. Mills closed.
Banks called in loans, and markets
filled with grown men, bagging
groceries, shopping while their wives
worked. When time is all you have
yards are manicured, cars repaired
houses painted one last bright time.

They told us the Japanese did it
so we threw rocks at foreign cars.
Up in the Pittsburgh office
they bought oil companies
and real estate. We didn't
understand business either.

For years they chanted Recovery.
It sounded like a prayer. We waited.
When they said clear out your locker
and the For Sale sign went up,
we understood.

Last month someone voted Pittsburgh
the best city to live in. Someone
working somewhere else. Things
are easier to see in this new light.

History is our life now.
Like scholars, we are the subjects
of our own idle debate.

A NOTE ON SOURCES

This book is primarily based on interviews with former steelworkers, and in some cases their widows and children. Unpublished materials I used include company catalogs, monographs, pamphlets, photos, and memorabilia in the possession of those steelworkers.

I consulted many newspapers, magazines and Web sites, such as Wikepedia while preparing this manuscript—too many to list here; however, I would be remiss if I did not mention how much I benefited from published articles in the *Union-Sun & Journal*. Some of these I found in the archives at the office of the Niagara County Historian, the Niagara Historical Society, a search of the microfiche records at the Lockport Public Library, and at Fulton History.com. Other articles were in the possession of the steelworkers who shared them with me.

I do not present this book as an academic work. It is neither footnoted nor extensively referenced. However, I drew on several excellent sources for the history of steel in general, and steelmaking in Western New York. Of those sources, which are listed in the selected bibliography, the book closest to mine, in terms of a steel industry memoir, if not geography, is Deborah Rudacille's *Roots of Steel*.

SELECTED BIBLIOGRAPHY

Bell, Thomas. *Out of this Furnace. A Novel of Immigrant Labor in America.* Little, Brown and Company, Pittsburgh, 1941.

Campbell, Rick. *History of Steel. A Selected Works.* All Nations Press. White Marsh, Va., 2014.

Diamond, Jared. *Guns, Germs and Steel.* W.W. Norton & Co. 1997.

Egan, Dan. *Death and Life of the Great Lakes.* W.W. Norton & Co. New York, 2017.

Eisler, Peter. "Poisoned Workers and Poisoned Places. Toxic Exposure Kept Secret." *USA Today*, September 6, 2000.

Flynn, Michael. "A Debt Long Overdue." *Bulletin of the Atomic Scientist.* 57; 2001. pp. 38–48.

Forging America: *The Story of Bethlehem Steel. "The Morning Call."* Tribune publishing Co., 2010.

Goldman, Mark. *Albright. The Life and Times of John J. Albright.* Buffalo Heritage Press, 2017.

Hoerr, John. *And the Wolf Finally Came.* University of Pittsburgh Press, Pittsburgh, 1988.

Kean, Sam. *The Disappearing Spoon. And Other Tales. . . . from the Periodic Table of Elements.* Back Bay Books/Little, Brown and Company. New York, 2010.

Kelly, Jack. *Heaven's Ditch. God, Gold and Murder on the Erie Canal.* St. Martin's Press, New York, 2016.

Korda, Michael. *Ike: An American Hero.* HarperCollins, New York. 2007.

Metzger, Jack. "Lackawanna & Johnstown: Shutdowns, Steel Towns and the Union." *Labor Research Review.* Cornell University ILR School, 1983.

McCullough, David. *The Great Bridge.* Simon and Schuster, New York. 1972.

Morgan, Spencer. *Western New York Steel.* Arcadia Publishing, Charleston, 2014.

Reutter, Mark. *Making Steel: Sparrows Point and the Rise and Ruin of American Industrial Might.* University of Illinois Press, Urbana, 2004.

Riley, Kathleen. *Lockport: Historic Jewel on the Erie Canal.* Arcadia Press, Charleston, 2005.

Regovin, Milton, and Frisch, Michael, *Portraits in Steel.* Cornell University Press, 1993.

Roberts-Abel, Catherine. *Simonds Saw and Steel, the Atomic Weapons Program and the "Myth of Practically Innocuous Radiation."* Thesis. State University of New York, Empire State College, 2003.

Rudacille, Deborah. *Roots of Steel. Boom and Bust in an American Mill Town.* Pantheon, New York, 2010.

Southard, Susan, Nagasaki. Life after Nuclear War. Penquin/Random House, New York, 2015.

Stoddard, Brooke C. Steel: *From Mine to Mill, the Metal that Made America.* Zenith Press, Minneapolis, 2015.

Strohmeyer, John. *Crisis in Bethlehem: Big Steel's Struggle to Survive.* University of Pittsburgh Press, Pittsburgh, 1994.

Terrill, Marshall. *Steve McQueen: The Life and Legend of a Hollywood Icon*. Triumph Books, Chicago, 2010.

Vance, J. D. *Hillbilly Elegy*. HarperCollins, New York, 2017.

Venetsky, S. *Tales About Metals*. Mir Publishing, Moscow, 1981.

Verhoeven, John. *Metallurgy for the Non-metallurgist*. ASM International, 2007.

Waldman, Jonathan. *Rust. The Longest War*. Simon and Schuster, New York, 2015.

GLOSSARY

Alloy   A mixture of elements. Basic steel is a mixture of iron (Fe) and carbon (C). By adding other elements, such as cobalt (Co), basic steel becomes uniquely designed for a wide range of applications. See supplemental appendix for alloying elements of Simonds steel.

Annealing   Heat treatment to soften steel to increase its workability. Needed after hot rolling to flatten/straighten and cold rolling to offset hardening process. Speed of heating and cooling adjusted for each product.

AOD   Argon-oxygen-decarburization process done in a special vessel after initial melting in the electric furnace. A three-step enhancement to inject oxygen and argon gas to remove carbon and sulfur and recover chromium.

Austenitic   The grain or crystalline structure of a mostly non-magnetic class of stainless steel that contains different percentages of nickel and chromium and sometimes molybdenum. Differs from ferritic (ferromagnetic) stainless steel.

Billet   Formed from an ingot that is heated to soften and reduce in size in a forge (a hammer or press). The billet is then sent for hot rolling. Grain size begins to get "finer" by the deformation.

Bi-metal   See Thermostatic.

Brinnell   Common test for hardness using a small press to form an indenture whose diameter is measured with a microscope's ocular micrometer.

Butt Weld   Welding method used to join ends of hot rolled thermostatic coils together to form longer coils before cold rolling.

Carbon   One of two essential chemical elements in standard steel varying from 0.2 to 2.1% as an alloying element with iron. Standard high-carbon steel is hard steel with minimal if any alloying elements. As the carbon percentage is decreased the working hardness decreases, but added alloys can recover hardness and impart special characteristics.

Cast Magnet   The foundry or casting department at Simonds where various types of permanent magnet and soft magnet alloys were molded in various forms to take advantage of magnetism's ability to hold ferromagnetic material and direct electron flow. The activities included casting design, sand casting, induction furnaces, special heat treatment furnaces and engineering support in the Met. Lab.

Cold Roll   The process to roll steel through special mills at room temperature (below the recrystallization temperature of a steel alloy) to alter the thickness of the sheet and strip and impart a shiny hard surface. At Simonds the cold roll department included "2 High mills," "4 High mills" and Sandzimir mills (aka "Z-mills.") The cold roll building also housed annealing furnaces, butt welder, slitters and cut-off machines.

Crucible   The original method used to melt steel at Simonds in the early 1900s. Refers to high-heat-resistant clay-graphite containers (crucibles) used in induction melting (see Induction.) Phased out in favor of electric arc method.

Decarburize   Removal of carbon when steel is melted or heated. This occurs at initial melting to affect carbon level throughout the cross section and also after melting to affect some degree of surface depth only. See AOD.

Electric Arc   One of several types of furnaces (See Furnaces) used in the steel industry to make, melt or reheat steel. The electric arc (EAF) uses an arc generated by carbon electrodes in direct contact with scrap to melt the metal to a molten state. A key element is the transformer, which steps down the voltage from the utility provider through the furnace's intricate circuitry delivering power in kilowatt hours converting scrap to steel.

EB Weld (Electron Beam Welding)   A method and apparatus to accelerate a beam of electrons in a vacuum at high speed to join metals together. At Simonds, used exclusively on the manganese alloy tri-metal product. (See Thermostatic.)

Ferritic   The grain or crystalline structure of a magnetic class of stainless steel of nearly pure iron, mainly alloyed with chromium.

Forging   Slow heating of ingots followed by hammering or pressing at temperature to form billets. At Simonds the hydraulic press replaced the steam hammer.

Flattener   Device used after hot rolling to improve the shape of steel by leveling under pressure. Straightener device in a similar category used for bar mill products.

Furnaces   Various types include blast, open hearth, induction and electric arc for melting. Oil, gas-fired, and electric used for heating before forging/rolling and heat treating. Heat treatment furnaces located throughout the plant; often referred to as "annealing furnaces." Also see Electric arc, Induction, and VIM.

Grains   The microscopic structure of steel; crystalline latticework composed of atoms with grain boundaries that are characteristic of each alloy. The grain size and distribution may be altered by heating, hot working, and cold working. Results checked microscopically for quality control in the Met. Lab where image analysis software has replaced the human eye.

Grinders Abrasive wheels used in different apparatuses; types included swing-frame and semi-automatic slab/billet. There were also hand grinders to remove small defects and surface grinders to do close-tolerance work. The principal tools of the cleaning (grinding) department at Simonds.

Hammer Device to forge hot steel by tapping/hammering. The original steam-driven forge at Simonds was replaced by 2,000-ton hydraulic press.

Hardening A type of heat treatment; reheating steel to high temperature followed by quenching and tempering to impart hardness or strength.

Heat Number A specific "heat number" assigned to any material melted. This identification follows the material until final shipment to the customer. The number is important for certification and quality control.

Hot Top Frame holding refractory material such as asbestos-containing mortar placed on top of hot ingot to trap impurities that rise to the surface as the ingot cools. Alumino-silicate, magnesium and other elements have replaced asbestos in hot tops. (See Refractory.)

Hot Rolling Passing hot forged steel bars or slabs above their crystallization temperature thorough one or more pairs of rolls to reduce thickness and width. There are many types of rolling processes and mills. At Simonds "HRA," hot rolled annealed only, was a common order for steel bars. After annealing they were straightened and cut to order. "HRAP" was a common order for sheets. After annealing they were pickled, flattened and cut or circle-punched to order.

Inclusion Defect in steel removed by chipping or grinding to provide a smooth workable surface.

Induction A basic type of furnace for melting steel where a crucible surrounded by electric oils melts the steel. Molten material is then poured into ingot or sand molds. A newer process, called Vacuum Induction Melting (VIM).

Ingot The form of molten steel poured into a mold. After cooling, the solidified ingot is removed and sent for further processing in a hammer or press to be made into billets. The grain size is very large in an ingot, still large in a billet, but progressively smaller with subsequent processing into slabs, sheets and bars. In some steel mills an intermediate reheated step between ingot and billet is called a "bloom."

Iron (Fe) One of two essential chemical elements needed in basic steel as an alloying element with carbon. See alloy list.

Magnets See Cast Magnet.

Met. Lab Metallurgical Laboratory where the physical properties of alloys are tested for such attributes as size and grain structure, hardness, heat resistance, and tensile strength. At Simonds, this included facilities for destructive and non-destructive testing, a machine shop, chemistry lab, research and development, cast magnet engineering, quality control, and a robust certification program to accompany and support customer shipments.

Pickling Acid treatment of steel to remove scale. Includes steps with different types of acids and other liquids, depending on content of carbon and other alloys. Modern-day alternatives include "environmentally friendly" options.

Quenching Controlled cooling of hot steel in air or liquid; e.g., water, oil or brine to harden steel.

Refractory Heat-resistant ceramic products—oxides, carbides or nitrides of metals, e.g. alumina, magnesia, zirconia, used in the manufacture of bricks to line high-temperature furnaces.

Revert Scrap   A byproduct of steel manufacturing which includes the discards, ingots, trimmings and rejects of operations categorized by alloy content. A cleaner, better-quality scrap metal.

Re-melting   Secondary melting operation using the original melted ingot as an electrode to produce a second or third ingot with increased chemical purity; a super alloy for the high demands of aerospace industries. Electro-slag (ESR) and vacuum arc (VAR) are types of re-melt processes.

Rockwell   A test for hardness using a device to measure the depth of penetration of an indenter under a specified load.

SAW   Submerged arc welding process where the arc is buried in flux. At Simonds it was used to edge weld bi-metal and tri-metal billets before hot rolling.

Shears   A housing with an industrial blade to cut-to-length/width steel plates, sheets and bars. Block-out shears, a robust type, cut plates from cogging mill. Simonds replaced old style "buffalo" shears in finishing departments with a plunger-type device.

Stainless Steel   A steel alloy with a chromium content of at least 10.5%. It resists oxidation; i.e., rust or corrosion. However, stainless can degrade like all metals, especially under conditions of low oxygen, high salinity and moisture. *Stain less* is not necessarily *stain free*. Three major groups: austenitic, ferritic and martensitic, but more than 150 different grades. At Simonds dozens of 300 series austenitic and 400 series ferritic stainless steels were made and marketed.

Tempering   A type of heat treatment. Reheating steel above a critical temperature that varies with each alloy for a desired property such as toughness. Performed after the initial heat treatment and quenching.

Thermostatic   Unique steels that respond to temperature change. When bonded to other steels such as in Bi-metal and Tri-metal, the product's response to temperature is controlled expansion or "bending." The most common base alloy for bonding is 36% nickel (aka Invar). The most demanding thermostatic product was a manganese/copper/nickel alloy melted in the foundry.

Tool steel Alloys with distinctive features of hardness, resistance to abrasion and deformation; maintains cutting edge at high temperature. Dozens of standard and custom products with various chemical compositions made at Simonds. Tungsten, chromium, molybdenum, vanadium and cobalt are principle ingredients.

Tri-metal See Thermostatic.

VIM Vacuum induction melting. Melting scrap and alloys in the low atmospheric pressure of a vacuum provides a cleaner atmosphere to draw off impurities while also avoiding contaminants in factory air. Today the VIM is followed by remelting by ESR and/or VAR to produce ultra-clean steel for aerospace and nuclear industries.

SIMONDS STEEL ALLOYS

Carbon (C)   While not typically considered an alloy, it is a critical element in alloy design. It forms simple and complex carbides with the elements listed below, and its percentage determines the degree of hardness, brittleness and corrosion resistance of the alloy.

Chromium (Cr):   Gives steel hardness and wear, corrosion and oxidation resistance. Minimum of 10.5% defines stainless steel forming the film that shields the surface. An important component in tool steel; alloyed with nickel, cobalt, molybdenum, magnesium, and manganese for a wide range of steel with a multitude of purposes.

Cobalt (Co):   Principle component of magnetic steel. It's one of three ferromagnetic metals along with iron and nickel (the iron triad). Has the highest Curie point (temperature at which it becomes non-magnetic; nickel is lowest and iron is intermediate). A component of heat-resistant steels.

Iron (Fe):   The component in scrap melted with other metals. Steel is iron alloyed with carbon (basic carbon steel). Wrought iron, cast iron and steel are similar alloys differing in carbon content. Addition of other elements in this list creates the stainless and tool alloys.

Molybdenum (Mo):   Increases steel's hardness at high temperature and ductility. Key component in WWI and WWII armor steel.

Retards growth of grains, imparting fine grain quality. Has high alloying capacity with other metals (cobalt, chromium and nickel); gives acid resistance.

Magnesium (Mg): A modifier of molten iron, it helps to deoxidize steel and steel alloys. Used for strong, light-weight alloys with aluminum and other metals. Applications in aircraft, rockets, and nuclear reactors, engine parts and many other products.

Manganese (Mn): Self-hardening, deoxidation, desulfurization and dampening ("quieting") properties taken advantage of in variety of steel alloys including iron (ferromanganese). Used in tool steel where shock is present in forging and stamping operations and in tri-metal expansion alloy.

Nickel (Ni): Component of wide range of alloys from armor plate to surgical instruments. Part of iron triad with iron and cobalt, it is a significant component of meteorites and magnet steels like Alnico (aluminum, nickel, cobalt). Expansion steels like invar and bimetals take advantage of its "memory." Strong, ductile and corrosion resistant, it provides a lustrous finish.

Tungsten (W): Its symbol (W) derives from the German name of the metal, wolfram. It is the principle component of tool steel along with chromium and cobalt. Similar to molybdenum in hardness, it has the highest melting point and tensile strength of all metals. It also has excellent corrosion resistance.

Titanium (Ti): Important Space-Age metal, lighter but harder than iron. Non-magnetic and non-conductive, it retains strength at high temperature and has remarkable resistance to corrosion. In stainless alloys it reduces grain structure.

Vanadium (V): Gives steel hardness, toughness and resilience. Mechanical strength, corrosion resistance and high melting point make vanadium-containing steel alloys important to automotive, aeronautic industries.

SIMONDS SAW AND STEEL
LOCKPORT TIMELINE

1832   J.T. Farwell & Company established in Fitchburg, Mass

1851   A. Simonds & Company

1864   Simonds Brothers & Company (The Fab 5)

1894   Dan Simonds President

1894   Simonds Establishes Steel Mill in Chicago

1910   Simonds Manufacturing Relocates—Chicago to Lockport

1911   Tool Steel by Crucible Method—250 Employees

1916   Electric Arc Furnace (6 ton) Steel

1918   World War I—steel armor plate and helmets

1922   Simonds incorporates as Simonds Saw and Steel

1930   Stainless Steel added

1934   Cast Magnet Steel added

1942   World War II—Armor Plate

1943   USW Local 2857 Established

1946   Uranium and Thorium Steel Rods Rolled ('46–'54)

1954   Cold Roll Mill

1958   Forge Shop with Hydraulic Press

1959   New Press Shop with 15 ton Electric Furnaces

1960   Bar Mill finishing and Metallurgical Lab Expansion

1965   Simonds sold to Wallace-Murray

1968   Guterl Steel Acquires Simonds Division of Wallace-Murray

1980 $15 Million Government Loan—Vacuum Induction Melting Installed

1982   Guterl Bankruptcy; End of Operations

ACKNOWLEDGMENTS

The idea that shaped this book first came to me after my son, Michael, shared photos of the abandoned Simonds steel mill he had taken on our trip to Lockport for my mother's funeral in 2010. The images rekindled memories of my steelworker summer, which I had touched briefly on in a memoir I published that year about growing up in Lockport. The idea to delve deeper coincided with a writer's workshop on the art of interviewing at The Piper Center at Arizona State University—it seemed like the perfect opportunity to connect with former steelworkers. The initial workshop assignment was to record and transcribe an interview as the basis of a short story. That first interview was the nidus of the manuscript that became *Men of Steel;* my memories interwoven with those of other *real* steelworkers and the history of Simonds Saw and Steel. I only wish I could have interviewed my father, grandfathers, uncles and friends of that generation.

I could not have written this work without the help of all of those other men and women—steelworkers, spouses and children—who shared their stories with me. I am grateful to Dan Bangleman, Reggie Buri, Jim Calos, Henry Ciarfella, John Coleman, Dave Craine, Anthony D'Angelo, Virginia DaVoe, Bill DeCesare, Dave DeLang, Al Ferrante, Tom Fiegl, Shirley Hayden, Barry Hemphill, Harold Kinsler, Louie

Koel, Steve Lacki, John Linder, Burt Malcolm, Gordon Martin, Jim McCormick, Joe Murphy, Tom Nichols, Mike O'Donnell, Tony Parete, John Peace, Maryann Rankie, Anthony "Doc" Ruggeri, Gracie Scirto, Arlene Sweet, John "Skeeter" Spry, Adrian Sherman; and Lou Valery, who was my first interview. I knew how important those stories are to each person I interviewed. Most of these people loved their jobs and grieved their loss. I could see it in their eyes and hear it in their voices. It even came through in email messages. Memories, as we know, are not infallible, and two or more steelworkers remembered the same event differently more than once in my interviews. I have tried to reconcile the differences. The written dialog is realistic although not always verbatim. I may have placed a word or two in their mouths they did not utter, but I didn't make up any of the exchanges.

I was given many names to contact. I reached many but not all of them, and I used most of their stories, sometimes combining similar experiences. There are undoubtedly many more retired steelworkers out there who would have contributed. I am sorry for having missed them and for those who I may have forgotten to thank.

I received much help from the archives at Niagara County Historian Office and the Niagara County Historical Society. Catherine Roberts-Abel's thesis, which I discovered at the Niagara Historical Society just a few paces down Niagara Street from where I grew up, provided a great start on the history of Simonds and the radiation issue. The reference librarian and staff at the Lockport Public Library gave me access to their Simonds articles and microfiche records of the *Union-Sun & Journal*. I thank the managing editor at the *Union-Sun*, Joyce Myles, for publicizing my project which prompted steelworkers to contact me, and for her helpful suggestions. The newspaper itself was the source of numerous articles; some of these articles came from the personal collections of steelworkers, especially Louie Koel to whom I am grateful. I am also indebted to Jim Calos, John Coleman, Burt Malcolm, Gordon Martin and Steve Lacki for the loan of catalogs, pamphlets, books, photos and memorabilia. I must identify some of these long-out-of-print items, as they provided specific information about Simonds that would have not otherwise have come to my attention. Thanks to John Coleman for loaning me his long-out-of-print copies of *Simonds Manufacturing Company's Seventy Five years of Business Progress and Industrial Advance, 1832-1907*, originally published by University Press, Cambridge, as well as *125 Years of Growth, 1832–1957*, published by the Simonds Saw and Steel Company, and finally *Simonds Steel catalog 121 and Useful Reference Tables* published by Simonds Manufacturing Company, 1921. *A Dedication to Excellence and Innovation. The Simonds Story 1832 to 2007*, is an update of Simonds' first seventy-five years. Many thanks to Ray Martino, President and CEO of Simonds International, for that booklet and permission to use images which allowed me to come full circle from Simonds' origins to the present time.

The librarians and archivists in the Special Collections at the Pennsylvania State University Library provided key records about the formation of the Steelworkers Union Local 2857 and early business items that were copied to United Steelworkers of America, District 4 office in Buffalo. (Three specific folders; 13, 14, and 15, pertaining to Simonds for the years from 1943–1990, were culled from Box 20 of the Labor Archives, A.1502.) Thanks to labor archivists Jim Quigley, Rachel Dreyer and Alexandra Arginteanu, without whom I could not have written the chapter on unionization. A visit to The Steel Plant Museum of Western New York in Buffalo introduced me to exhibits and historical information about the area's steel industry and steelworkers who labored in the plants, although there was little information about the Simonds operation in Lockport.

I thank Craig Bacon for reviewing the manuscript and offering suggestions. David Harmon, a friend and ceramic engineer rectified a misinterpretation I had regarding furnace linings, and Carl Schneider's explanation of the Ferromagnetic Hysteresis Loop corrected a misunderstanding about magnetic steel. Any technical errors that remain are my responsibility alone.

I am most appreciative to Craig Rowmanoski at the Allegheny Ludlum ATI facility in Lockport, and Bob Shabala and Tim Getter at Niagara Specialty Metals in Akron for the time they spent explaining their operations, the tours of their facilities, vintage photos and the reference materials. Special thanks to Rick Campbell for his permission to excerpt his entire poem, History of Steel, from his recently published collection, which captures sentiments I attempted to convey in my manuscript.

I thank my writing mentors: Katie Grant, Marshall Terrill, and Susan Southard, who provided suggestions and guidance. My friends Victor Amoroso, Cesar Berardi, Dominick Dellaccio, Anthony Farone, Roy Pipitone, Frank Rinaldo, and Jim Sansone assisted me with local research; and I thank David Wallace for his assistance with the cover concept, and the initial arrangement of the vintage and abandonment photo sections uploaded to 1106 Design.

I'm indebted of course to my son, Michael, who contributed more than his photographs—only a select few were used in this book. Light, shadow and colors, particularly the many hues of rust and foliage, are difficult to convey in small black and white images. I look forward to the publication of his monograph, *Moldering*, with its many striking images of urban and industrial decay.

This book could not have been published without the work of the staff at 1106 Design. Special thanks to Ronda—"Team work makes Dream work."

Finally I thank my wife, Rosalie, who read the initial and final drafts and offered me the suggestions and support I needed to persevere with this project.

ABOUT THE AUTHOR

D r. Louis A Rosati was born and raised in Lockport, New York where he attended Lockport public schools. He graduated from the University of Buffalo ('62) and the Upstate Medical Center at Syracuse ('66). He did his residency in pathology at the University of Michigan Medical Center in Ann Arbor, and then served in the Navy at the National Naval Medical Center in Bethesda, MD. He practiced pathology for 37 years in the Phoenix metropolitan area where he was a co-founder of Clin-Path Associates and Sonora Laboratory Sciences (Sonora-Quest). Now retired, Louis resides in Mesa, Arizona with his wife Rosalie, his Lockport High School class of '58 sweetheart of more than 55 years. His publications include pathology articles and book chapters in the peer-reviewed medical literature, and a creative non-fiction book—*My Winning Season*, which traces the summer of 1954, in a memoir about growing up and playing baseball in Lockport.

The author poses before his father's sheet mill at Niagara Specialty Metals, Akron, New York

www.ingramcontent.com/pod-product-compliance
Lightning Source LLC
Chambersburg PA
CBHW061628220326
41598CB00026BA/3923